Cyber Power Potential of the Army's Reserve Component

Isaac R. Porche III, Caolionn O'Connell, John S. Davis II, Bradley Wilson, Chad C. Serena, Tracy C. McCausland, Erin-Elizabeth Johnson, Brian D. Wisniewski, Michael Vasseur

Prepared for the United States Army
Approved for public release; distribution unlimited

For more information on this publication, visit www.rand.org/t/RR1490

Library of Congress Cataloging-in-Publication Data is available for this publication.

ISBN: 978-0-8330-9480-3

Published by the RAND Corporation, Santa Monica, Calif.

Front cover photo by Sgt. Stephanie A. Hargett/U.S. Army

Preface

This report documents research conducted as part of a study entitled "Managing and Developing Reserve Component Capabilities in Support of the Army's Cyber Force." The primary purpose of the study was to conduct initial research on how to train, manage, and develop the Army's cyber force, with a specific focus on the Army National Guard and the U.S. Army Reserve.

In this report, we describe the availability of personnel with cyber skills in the private sector and the number of "citizen-soldiers" available to support the Army's cyber mission needs. This report will be of interest to the entire reserve component, Congress, Army Cyber Command, and U.S. Cyber Command.

This research was sponsored by the U.S. Army National Guard; the Office of the Chief, Army Reserve; and the U.S. Army Cyber Center of Excellence and conducted within the RAND Arroyo Center's Personnel, Training, and Health Program. RAND Arroyo Center, part of the RAND Corporation, is a federally funded research and development center sponsored by the United States Army.

The Project Unique Identification Code (PUIC) for the project that produced this document is HQD156884.

Executive Summary

The military services are formalizing and bolstering their contribution to the nation's cyber force, known as the U.S. Cyber Command Cyber Mission Force. As a part of a Total Force approach, the Army is considering using both active component and reserve component (RC) personnel to fill the Cyber Mission Force and other requirements in support of Army units.

This report identifies the number of Army RC personnel with cyber skills, to help identify ways in which these soldiers can be leveraged to conduct Army cyber operations. This report also describes the broader challenges and opportunities that the use of RC personnel presents.

To study this issue, we first performed a thorough review of past studies, government reports, and relevant literature. Next, we analyzed data from the Civilian Employment Information database and the Work Experience File database. We performed analyses of social media data from LinkedIn profiles, which include self-reported cyber skills among reservists. Also, we reviewed and assessed the knowledge, skills, and abilities (KSAs) defined for Cyber Mission Force roles in order to determine the percentage of these KSAs that can be acquired in the private sector. Finally, we conducted of a survey of more than 1,200 guardsmen and reservists.

Based on both quantitative and qualitative analyses, we find that relevant information technology and cyber skills are in abundance in the private sector. As a result, there are tens of thousands of "citizen-soldiers"—that is, soldiers in the Army RC—who have the potential to support the Army's cyber mission needs or the propensity to learn cyber skills.

Contents

Summary

As threats and opportunities in the cyber domain increase, the military services are formalizing and bolstering their contribution to the nation's cyber force, known as the U.S. Cyber Command Cyber Mission Force. The Army is working to acquire, train, manage, and develop cyber capabilities, and, as a part of a Total Force approach, the Army is considering using both active component (AC) and reserve component (RC) personnel to fill the Cyber Mission Force and other requirements in support of Army units.[1]

In fiscal year 2015, the RAND Arroyo Center embarked on a study entitled "Managing and Developing Reserve Component Capabilities in Support of the Army's Cyber Force." As part of this study, we were asked to inventory the cyber skills resident in the Army's RC and identify possible cyber roles and missions for the RC. We were also tasked to recommend approaches to recruiting, training, and assigning RC cyber personnel to support Army cyber operations.

Approach

We employed a number of methods to achieve these objectives:

- a thorough review of past studies, government reports, and relevant literature

[1] For the purpose of this report, the terms *reserve component* and *RC* encompass both Army National Guard and U.S. Army Reserve forces.

- observations at cyber conferences and exercises (e.g., CyberGuard 14, CyberFlag)
- analysis of data from the Civilian Employment Information (CEI) database and the Work Experience File (WEX) database
- analysis of social media data from LinkedIn profiles
- a review and assessment of knowledge, skills, and abilities (KSAs) defined for Cyber Mission Force roles
- development, administration, and analysis of a survey of more than 1,200 uniformed personnel in the Army National Guard (ARNG) and U.S. Army Reserve (USAR).

Findings

Key findings from these efforts are enumerated in Table S.1.

Advocates for use of the RC in cyber operations cite a number of reasons. Chief among these is the RC's ability to provide surge capacity for various cyber roles within each service and for the Cyber Mission Force. In addition, the RC has been suggested as a means of retaining valuable cyber personnel, e.g., recouping the U.S. Department of Defense's (DoD's) investment in its extensively trained personnel when they leave the AC. For the National Guard in particular, a homeland defense mission is envisioned as an ideal role, especially given the increasing concern over the risk to the nation's critical infrastructure (including the power grid).

Pessimistic assessments of the value of the RC for the cyber domain are influenced by the lengthy training requirements in place today for key roles in the Cyber Mission Force and the possible unavailability of RC personnel to complete this training. However, it remains to be studied whether the DoD training and education regimen needs to include systems other than those most unique to the military (e.g., weapons). It is possible that civilian-acquired training and experience are already sufficient for a significant number of roles in the Cyber Mission Force. Of course, civilian-acquired training alone will not be adequate to prepare individuals for all, or even most, of the work roles associated with the Cyber Mission Force. But there are certainly many

Table S.1
Study Findings

#	Finding	Method
1	The Army will likely need more cybersecurity personnel in the future than it has today. This projected shortage is exacerbated by a rapidly growing demand for cybersecurity personnel in the private sector.	Literature review; RAND Arroyo Center analysis
2	The level of cyber expertise that exists in the RC can be estimated with currently available data sources, including the CEI database, the WEX database, and, potentially, novel uses of social media, such as LinkedIn profiles.	Analysis of CEI data and LinkedIn profiles
3	DoD and the Army would benefit from a more detailed inventory of their cyber professionals, relative to what is provided by the current CEI/WEX data.	Literature review; RAND Arroyo Center analysis
4	Most (but not all) of the KSAs needed for cyber operations—specifically, those identified by the U.S. Cyber Command as requirements for many of the roles that support the Cyber Mission Force—can be "civilian-acquired" via civilian-based training and experiences. Specifically, they can be acquired in part from popular certificate programs (e.g., Certified Ethical Hacker [CEH], Certified Information Systems Security Professional [CISSP], Security+) and civilian-sector on-the-job training.	Analysis of Cyber Mission Force KSAs; literature review; analysis of Cyber Mission Force tasks; analysis of RAND Arroyo Center survey data
5	Sufficient operations tempo is vital to stay "cyber-sharp." Many guard and reserve soldiers are employed in leading-edge technology companies and have critical skills and experience in fielding the latest information technology (IT) systems, networks, and cybersecurity protocols. Arguably, their nonmilitary employment allows them to more easily maintain currency in their cyber skills, compared with some AC soldiers who are not engaged in cyber tasks on a frequent basis.	Literature review; analysis of LinkedIn profiles
6	There are personnel in the RC whose civilian cyber expertise is not being utilized in or applied to their Army careers. This possible untapped cyber potential is approximately 11,000 people who, at a minimum, have the propensity to learn the cyber skills needed for Army cyber operations.	Analysis of the CEI/WEX database and survey data
7	There are strong indications that many in the pool of untapped cyber potential have a desire to use their cyber-related skills in the Army. Many others who do not have cyber skills have a strong interest in acquiring them.	Analysis of survey data

Table S.1—continued

#	Finding	Method
8	The Army will need to continually adjust its strategies for recruiting, training, and qualifying cyber specialists. Potentially effective options for reserve recruiting include the use of expanded age ranges and generous compensation for sufficiently trained personnel in the private sector.	Literature review; interviews with exercise participants
9	The Army should use a cyber aptitude assessment tool, similar to what the Air Force, the National Security Agency, and other countries utilize, to aid recruiting for cyber personnel.	Arroyo Center analysis

roles for which civilian-acquired training primes the individual to be trained to the level of expertise needed for active duty in the Cyber Mission Force.

Based on both quantitative and qualitative analyses, we find that relevant IT and cyber skills are in abundance in the private sector. As a result, there are tens of thousands of "citizen-soldiers"—that is, soldiers in the Army RC—who have the potential to support the Army's cyber mission needs.

Recommendations

Proceed with the Incorporation of RC Personnel into Plans for the Army's Cyber Force

Individuals in the ARNG and USAR whose IT training and experience has been enhanced by their civilian employment are ideal sources of cyber talent. There is sufficient overlap between the KSAs required for the Cyber Mission Force and those used in the civilian IT industry to suggest that there is value in the pool of talent employed there. The Army should leverage this pool to the maximum extent possible. At one time, the Air Force set up reserve units near Redmond, Washington, to take advantage of talent working for such IT-focused companies as Microsoft and others in that state. The Army has done the same. For example, the Army Reserve Cyber Operations Group has had subordinate elements aligned with technology centers since its inception. More advanced concepts, such as a "civilian cyber corps," make sense but, to a certain extent, can be achieved using the ARNG and USAR forces today. Furthermore, some roles that are offensive in nature demand uniformed personnel, especially if a presence in a theater of operation is required.

Increase Compliance with and Revise the CEI Questionnaire

We recommend that DoD find ways to increase the compliance with the annual CEI questionnaire, perhaps by issuing more frequent reminders to RC personnel regarding this mandatory task. It has the potential to be a great source of data and yield updated analyses on the cyber skills resident in the RC. The CEI questionnaire should also be modified to ask for greater detail with respect to cyber-related skills.

Develop a New Strategy to Manage the Future Cyber Workforce

Army Human Resource Command will need different processes and technologies than are used today to manage the cyber workforce. The cyber workforce will include new and emerging specialties and function areas, and equivalencies for real-world experiences will need to be continually examined and granted.

Acknowledgments

We benefited from the guidance and input of our sponsor representatives, including COL Aida T. Borras, COL Michael Vargas, Todd Boudreau, and MAJ Jody Wright.

We thank Jamaal Lockett, Scott Seggerman, and Tori Rodrigues of the Defense Manpower Data Center (DMDC) for making this analysis possible. We also gratefully acknowledge the following RAND colleagues: Suzy Adler for assistance in accessing the DMDC data; Anthony Rosello and Ellen Pint for assistance in understanding the Civilian Employment Information data set; Greg Schumacher for his insights on the RC; Pete Schirmer and Bruce Held for helping to develop and shape the study; Tom Curley for his comments on the U.S. Cyber Command; Ian Cook and Michelle Ziegler for contributions to discussions about private-sector IT staffing and cybersecurity needs; and Meg Harrell, Bryan Hallmark, and Heather Krull for their overall feedback on the research effort. Susan Straus provided guidance on issues related to human subjects and survey techniques. Christina Panis served as our research programmer, organizing data received from the DMDC. Kristin Van Abel provided geographical analysis of the CEI data and the illustrations of this analysis. We especially thank Michael Hansen for his leadership and guidance throughout the study effort and his careful reading and editing of this final report. We are also especially appreciative of Michelle McMullen, who served as our administrative assistant. Michelle reviewed and edited numerous versions of this report, organized the survey data we collected, and developed daily updates for the charts in this report based on that data.

There are a number of general officers in the Army who were instrumental in helping us with this research effort. MG Stephen G. Fogarty helped us disseminate our survey. BG Walter E. Fountain provided early guidance and instruction at the beginning of the study. BG Christopher J. Petty provided input as well. LTG Jeffrey W. Talley provided a number of constructive comments and suggestions on our final briefing that helped this overall effort. LTG Edward C. Cardon provided valuable and important comments on our written report.

Finally, we thank our reviewers, Cynthia Dion-Schwarz of RAND, LTG(Ret) Rhett A. Hernandez, and COL(Ret) Ralph Wayne Dudding. Their cyber expertise was invaluable in revising our earlier drafts. We remain grateful.

CHAPTER ONE

Introduction

The military services are formalizing and bolstering their contribution to the nation's cyber force, known as the U.S. Cyber Command Cyber Mission Force.[1] Within the Army, there are plans to use both active component (AC) and reserve component (RC) units in the Cyber Mission Force as part of a Total Force approach.[2] This report describes some of the challenges and opportunities that such an approach presents.

---

[1] U.S. Cyber Command has three primary missions: (1) secure, operate, and defend U.S. Department of Defense (DoD) networks; (2) defend the nation in cyberspace; and (3) support combatant command full-spectrum operations in cyberspace. These missions are to be carried out in part by the new Cyber Mission Force. In December 2012, DoD approved a plan to establish this new cyber force resourced from all of the services aligned to these missions. Implementation of the approved Cyber Mission Force plan is under way, with progress measured and reported on a quarterly basis. See DoD, *The DoD Cyber Strategy*, Washington, D.C., April 2015.

The Cyber Mission Force is composed of 133 teams. These teams are commissioned to execute the three missions listed above and are expected to be fully manned, trained, and equipped by fiscal year 2018. The Army will provide 41 of the teams (including the Cyber Protection Team [CPT]). The Army plans to draw from both the AC and RC and has initiated an analysis to develop a Total Army RC cyber integration strategy to support its requirements. See Edward C. Cardon, "2014 Green Book: Army Cyber Command and Second Army," web page, September 30, 2014.

[2] For the purpose of this report, the terms *reserve component* and *RC* encompass both Army National Guard and U.S. Army Reserve forces.

Background and Motivation

The RAND Arroyo Center was asked by the Army National Guard, the Office of Chief of Army Reserve, and the Army Cyber Center of Excellence to examine the cyber skills that exist in the RC. Motivating questions addressed include the following: How can the RC be leveraged for Army cyber operations? What is the right AC/RC skill mix for the Army cyber operations?

Current Usage of RC Personnel

According to March 2015 testimony from the current commanding general of Army Cyber Command, LTG Edward C. Cardon, "Army Guard and Reserve forces routinely [augment their] headquarters."[3] In addition, the RC continues to support current operations in Southwest Asia (e.g., the Regional Computer Emergency Response Team [RCERT]–Southwest Asia [SWA]).[4] Already, "the Army has activated a National Guard Cyber protection team in Title 10 status supporting ARCYBER [U.S. Army Cyber Command] and 2nd Army."[5]

U.S. Army Cyber Command's Planned Strategy

LTG Cardon's testimony outlines the following objectives of an integration strategy:

- building an operational reserve in the [RC] for cyberspace crisis response[6]
- dual-use capability in support of military and homeland defense and DSCA [defense support of civil authorities] missions
- organizing cyber units to match Cyber Mission Force structure

---

[3] Edward C. Cardon, "Operationalizing Cyberspace for the Services," testimony before the House Armed Services Committee Subcommittee on Emerging Threats and Capabilities, Washington, D.C., March 4, 2015.

[4] Cardon, 2015.

[5] Cardon, 2015.

[6] This corresponds to the current DoD cyber strategy of drawing on the National Guard and Reserve. The National Guard and Reserve represent the "DoD's critical surge capacity for cyber responders" (DoD, 2015).

- aligning [RC] cyber forces under ARCYBER [U.S. Army Cyber Command] training and readiness authority and leveraging industry connected skills and using the reserve components retention advantages.[7]

Future Usage of RC Units

According to LTG Cardon's testimony, a plan is in place to include "21 reserve component cyber protection teams trained to the same standards as the active component cyber force."[8]

Report Objective

The objective of this report is to inform the Army's effort to acquire, train, manage, and develop the Army's cyber force, with a specific focus on the utility of the Army National Guard (ARNG) and the U.S. Army Reserve (USAR). Results reported are intended to contribute to the development and refinement of a comprehensive human capital strategy for the Army's cyber force, ideally with a Total Force approach that integrates all the components. As LTG Cardon stated in his March 2015 testimony,

> Army Cyber Command is a total multi-component force of Active and Reserve Components which are fully integrated into the cyberspace force mix. Building the U.S. Army Reserve (USAR) and Army National Guard (ARNG) cyber forces is a high priority for the Army and ARCYBER [U.S. Army Cyber Command].[9]

---

[7] Cardon, 2015.

[8] Cardon, 2015.

[9] Cardon, 2015.

Methodology

We employed a number of methods to achieve these objectives:

- a thorough review of past studies, government reports, and relevant literature
- analysis of data from the Civilian Employment Information (CEI) database and the Work Experience File (WEX) database
- analysis of social media data from LinkedIn profiles
- a review and assessment of knowledge, skills, and abilities (KSAs) defined for Cyber Mission Force roles
- development, administration, and analysis of a survey of over 1,200 guardsmen and reservists.

Organization of This Report

This report is organized as follows. Chapter Two discusses the demand for information security professionals. Chapter Three describes our literature review, outlining findings from recent studies. Chapter Four details the Army RC cyber inventory analysis. Chapter Five describes the role and importance of civilian certification and training. Chapter Six presents an analysis of data from LinkedIn. Chapter Seven presents results from a survey of members of the USAR and ARNG. Chapter Eight considers ideal roles and missions for Army cyber forces. Chapter Nine provides a review of the Army's human capital strategy by comparing existing Army plans with ideas from other military services, government agencies, and the private sector. Chapter Ten summarizes the overarching observations and study findings. Appendixes supply relevant data and details.

CHAPTER TWO

The Growing Demand for Information Security Professionals

In this chapter, we consider trends in the private sector in order to estimate future demands for information security professionals. For the most part, this chapter focuses on industry terms and definitions with regard to cybersecurity. As we note in the discussion at the end of the chapter, DoD has its own terms and definitions. Nevertheless, both DoD and the private sector are seeking some of the same information technology (IT) and cyber skills, and both are recruiting from the same workforce. So, all of the terms are relevant, at least in terms of understanding the demand for information security professionals.

Private-Sector Trends and Metrics Suggest Growth

Information security (infosec) is defined as

> protection of information systems against unauthorized access to or modification of information, whether in storage, processing or transit, and against the denial of service to authorized users, including those measures necessary to detect, document, and counter such threats.[1]

[1] National Security Agency, *National Information Systems Security (INFOSEC) Glossary*, Washington, D.C., NSTISSI No. 4009, September 2000, p. 30.

The term *infosec* is nearly synonymous with the terms *information assurance* and *cybersecurity*.[2] The National Initiative for Cybersecurity Education's National Cybersecurity Workforce Framework established a common taxonomy and lexicon for describing cybersecurity work and workers.[3] The framework identifies seven key areas and specific tasks associated with each area:

1. **Securely provision:** specialty areas responsible for conceptualizing, designing, and building secure IT systems (i.e., responsible for some aspect of systems development).
2. **Operate and maintain:** specialty areas responsible for providing support, administration, and maintenance necessary to ensure effective and efficient IT system performance and security.
3. **Protect and defend:** specialty areas responsible for identification, analysis, and mitigation of threats to internal IT systems or networks.
4. **Investigate:** specialty areas responsible for investigation of cyber events and/or crimes of IT systems, networks, and digital evidence.
5. **Collect and operate:** specialty areas responsible for specialized denial and deception operations and collection of cybersecurity information that may be used to develop intelligence.
6. **Analyze:** specialty areas responsible for highly specialized review and evaluation of incoming cybersecurity information to determine its usefulness for intelligence.
7. **Oversight and development:** specialty areas provide leadership, management, direction, and/or development and advocacy

---

[2] Department of Defense Instruction 8500.01, *Cybersecurity*, March 14, 2014, defines *cybersecurity* as

> Prevention of damage to, protection of, and restoration of computers, electronic communications systems, electronic communications services, wire communication, and electronic communication, including information contained therein, to ensure its availability, integrity, authentication, confidentiality, and nonrepudiation.

[3] National Initiative for Cybersecurity Education, *The National Cybersecurity Workforce Framework*, Washington, D.C.: National Institute of Standards and Technology, 2013.

so that individuals and organizations may effectively conduct cybersecurity work.[4]

Figure 2.1 lists these specialty areas and subareas and highlights the ones usually associated with the terms *infosec* and *IT.*

Figure 2.1
Components of "Cybersecurity Work" Described in National Initiative for Cybersecurity Education Framework

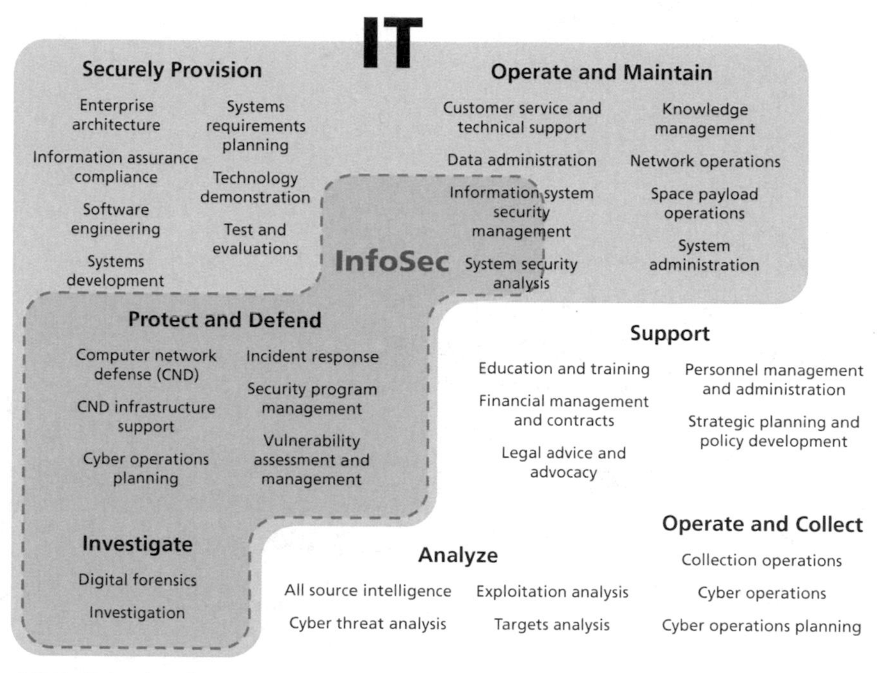

SOURCE: National Initiative for Cybersecurity Education, 2013.
RAND *RR1490-2.1*

[4] Paraphrased from National Initiative for Cybersecurity Education (2013).

Both Private and Public Sectors Claim to Be Understaffed

Surveys show that more than half of private- and public-sector organizations believe they do not have enough infosec personnel (Figure 2.2).[5]

The same survey suggests there is a broad breadth of knowledge and skills required for the success of infosec personnel (see factors listed in Figure 2.3).[6]

Figure 2.2
Does Your Organization Currently Have the Right Number of Information Security Workers?

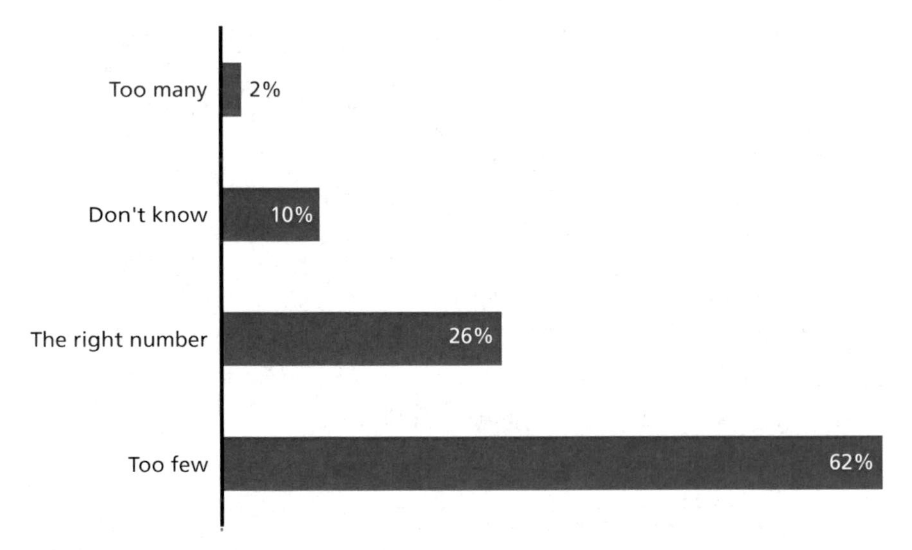

SOURCE: Data from Suby and Dickson, 2015.
RAND *RR1490-2.2*

---

[5] In *The 2015 (ISC)² Global Information Security Workforce Study*,

> the distribution by organization size spanned small (1–499 employees) at 25% of the survey respondents, mid-sized (500–9,999 employees) at 32%, and large at 43% . . . the 2015 survey was completed by 13,930 qualified information security professionals; a combination of (ISC)² members and non-members. (Michael Suby and Frank Dickson, *The 2015 (ISC)² Global Information Security Workforce Study*, Mountain View, Calif.: Frost & Sullivan, 2015)

[6] This chart does not indicate the role of training in these characteristics.

Figure 2.3
Success Factors for Information Security Professionals

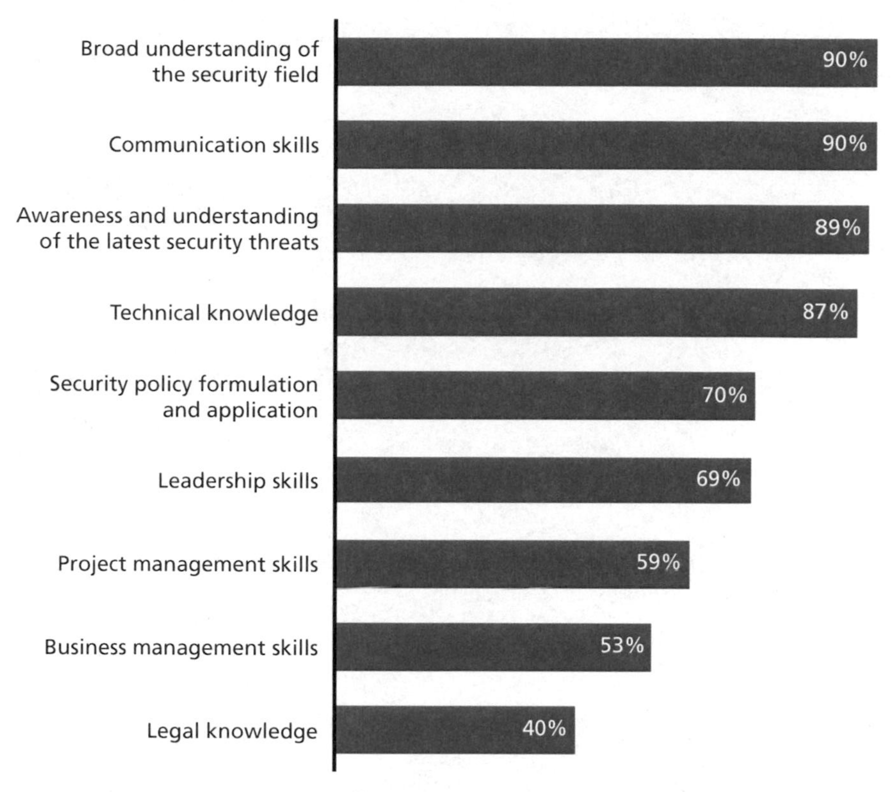

SOURCE: Data from Suby and Dickson, 2015.
RAND *RR1490-2.3*

Government projections show growth.[7] Specifically, the U.S. Bureau of Labor Statistics projects 37-percent growth in infosec analysts between 2012 and 2022, which is said to be "faster than average"

---

[7] This is a viewpoint shared by many, including John Klebonis, vice president of federal government customer service for AT&T: "Skilled cyber professionals are in high demand but short supply. The government and private sector will jockey for talent for the foreseeable future because of the dearth of qualified workers" (quoted in Sandra Jontz, "Uniting Cyber Defenses," *SIGNAL*, October 1, 2015).

(i.e., greater growth than the average for all occupations).[8] These projections are illustrated in Figure 2.4.

Figure 2.4
Information Security Analysts: Percentage Change in Employment, Projected 2012–2022

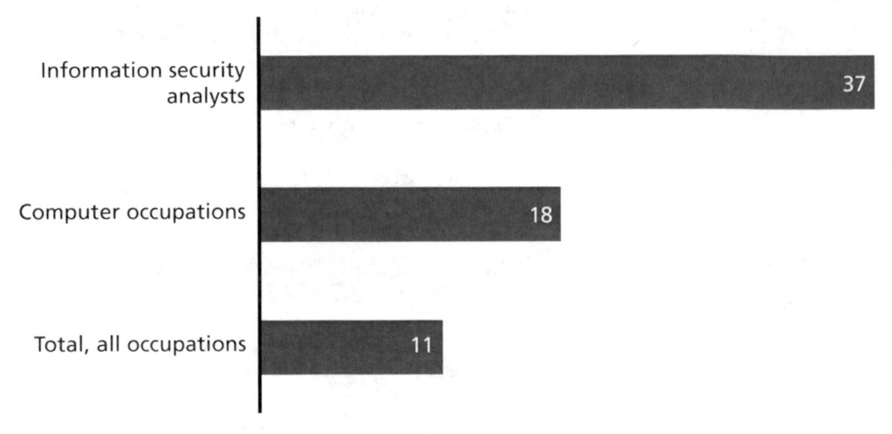

SOURCE: Data from U.S. Bureau of Labor Statistics, "Employment Projections," web page, undated-a.
NOTE: The term *all occupations* refers to all occupations in the U.S. economy.
RAND *RR1490-2.4*

Lessons from the Target Corporation Breach: People Matter

The 2013 breaches of Target Corporation, documented in a number of reports, highlight the role of human analysts and operators for infosec workers protecting networks and data.[9] Forty million credit and debit cards were compromised, and the personal data of more than 70 million shoppers were stolen. Bangalore-based analysts with the attack-detection firm FireEye noticed the attacks and notified the security team at Target headquarters in Minneapolis. However, insufficient

---

[8] U.S. Bureau of Labor Statistics, "Information Security Analysts: Summary," January 8, 2014b.

[9] For example, see Michael Riley, Ben Elgin, Dune Lawrence, and Carol Matlack, "Missed Alarms and 40 Million Stolen Credit Card Numbers: How Target Blew It," *Bloomberg Business*, March 13, 2014.

action was taken. In other words, it was not just tools that were to blame—there was a more significant "people failure."[10]

A Number of Metrics Are Used to Gauge Information Security Needs

Table 2.1 represents a summary of the literature estimating the number of personnel needed to perform infosec. There are varying metrics that are used to quantify the number of infosec personnel that are needed for a given organization or staff. Some efforts compare the number of infosec personnel with the number of users of computing resources in an organization. Others compare the number of infosec staff with the number of networked devices. Table 2.2 shows the ratio of IT staff to total employees for various organizational sizes. We note that, because of differences in how studies define IT personnel, the numbers in the table might not be entirely comparable.

The Increasing Ratio of Information Security Personnel to Total Staff

As Table 2.3 shows, infosec personnel constituted as much as 1.7 percent of all staff (as of 2011). On average, organizations were 0.53 percent infosec in 2011, compared with 0.06 percent in 1997 and 0.03 percent in 1987.[11]

The data provided in Table 2.3 are several years old and losing relevance in the age of cloud-based solutions. Nonetheless, even with older metrics, it is apparent that a significant number of infosec professionals are being employed in most sectors.[12]

---

[10] Jaikumar Vijayan, "Major Companies, Like Target, Often Fail to Act on Malware Alerts," *Computerworld*, March 14, 2014; Riley et al., 2014.

[11] Charles Cresson Wood, *Information Security and Data Privacy Staffing Levels: Benchmarking the Information Security Function*, Houston, Tex.: Information Shield, January 2012.

[12] According to a Ponemon Institute report, IT security department head counts are growing: "The average headcount of an IT security function is expected to grow from 22 staff members in 2013 to 29 in 2014" (Ponemon Institute LLC, *Understaffed and at Risk: Today's IT Security Department*, Traverse City, Mich.: February 2014, p. 12).

Table 2.1
Industry Considers Benchmark Ratios

Ratio	Benchmark	Description	Source
infosec:user	1:1,000	1 infosec staff per 1,000 users	Deloitte Touche Tohmatsu, 2003; Wood, 2012
Infosec:IT	3–5:100	3–5 infosec staff per 100 IT staff	Wood, 2012; Tipton and Krause, 2007
Infosec:IT	6–8.5:100	6–8.5 infosec staff per 100 IT staff	Aubuchon, 2010
Infosec:IT	1.5–2:100	1.5–2 infosec staff per 100 IT staff	"IT Security Staff Levels Are Declining," 2008
Infosec:IT	3–4:100	3–4 infosec staff per 100 IT staff	Kvavik et al., 2003
Infosec:auditor	1.75:1	1.75 infosec staff per internal IT auditor	Wood, 2012
Infosec:devices	1:5,000	1 infosec staff per 5,000 networked devices	Pirani, 2004
Infosec:IT	3–11:100	3–11% of overall IT budget allocated to infosec	Aubuchon, 2010; Vostrom Holdings, Inc., undated
Users:IT staff	20:1	20 other employees per IT staff member (operate and maintain; dependent on sector, defense contractor)	Ongoing Air Force study survey (Schmidt et al., 2015); dependent on sector
Infosec:IT staff	15–20:1	15–20 infosec staff per 1 IT staff member	Ongoing Air Force study survey (Schmidt et al., 2015); dependent on sector

Discussion and Observations

Staffing Is Difficult

Competing with high-tech companies (such as Google and Facebook) to recruit infosec professionals is difficult. These companies can provide their staff with high salaries and attractive perks.[13] Absent a major cultural shift, demand for these staff will grow, and competition will

[13] Philip Ewing, "Ash Carter's Appeal to Silicon Valley: We're 'Cool' Too," *Politico*, April 23, 2015.

Table 2.2
Ratio of IT Staff to Total Employees

Organization Size	25th Percentile	50th Percentile (median)	75th Percentile	Organization Count
All organizations	1:11	1:27	1:52	103
By Annual Dollar Volume				
Less than $200 million	1:11	1:19	1:34	25
$200 million to <$500 million	1:19	1:36	1:61	20
$500 million to <$1 billion	1:11	1:31	1:53	17
$1 billion to <$5 billion	1:20	1:36	1:82	20
$5 billion or more	1:10	1:15	1:25	20
By Total Number of Employees				
Fewer than 500	1:8	1:18	1:34	16
500 to <1,000	1:14	1:25	1:40	14
1,000 to <5,000	1:11	1:23	1:45	38
5,000 to <10,000	1:10	1:25	1:53	15
10,000 or more	1:23	1:40	1:112	20

SOURCES: Brian Richardson, "Improve Staffing Ratios," *ZDNet*, February 11, 2002; Joanne Cummings, "Your Life in the Virtualized Future" *NetworkWorld*, July 26, 2004; Quorum Technologies, Inc., "Case Study: Alameda County Medical Center," 2008; Robert L. Mitchell, "Enterprise Linux? Not So Fast," *ComputerWorld*, January 19, 2009; Lon D. Gowen, *Predicting Staffing Sizes for Maintaining Computer Networking Infrastructures*, McLean, Va.: MITRE Corporation, 2000.

become even more difficult.[14] Figure 2.5 shows industry metrics from past studies to determine how many infosec personnel companies need. Other information provided includes the number of internal IT auditors and overall IT budget allocated to infosec.

---

[14] For example, some have speculated that massive worker burnout at high-tech companies, such as Amazon and Microsoft, could shift preferences. In that case, DoD could focus on these "burnouts" (especially those with families) who are interested in a second career.

Table 2.3
Percentage of Information Security Staff, by Industry

Business Activity	Information Security Staff/ Total Staff (%)
Government (federal)	1.68
Health care	1.30
Services/consulting	0.78
High-tech	0.55
Average	0.53
Financial services	0.52
Utilities	0.37
Government (state and local)	0.29
Education	0.26
Manufacturing/production	0.26
Other	0.24
Telecommunications	0.21
Transportation/distribution	0.17
Retailing/wholesaling	0.09

SOURCE: Wood, 2012.

Software Developers Will Also Be Needed

Network defense techniques will require dynamic networks and dynamic development by individuals. Networks that are dynamic (e.g., software-defined networks,[15] polymorphic configurations) will require human defenders to be responsive and creative and thus able to develop new tools and scripts constantly. The increasing number of zero-day attacks (i.e., attacks using previously unknown approaches) will require

[15] Brian Roach, "3 Reasons Software-Defined Networking Is Streamlining DoD IT," *Defense Systems*, April 14, 2015.

Figure 2.5
Information Security Needs Vary by Sector, but the Overall Trend Is Upward

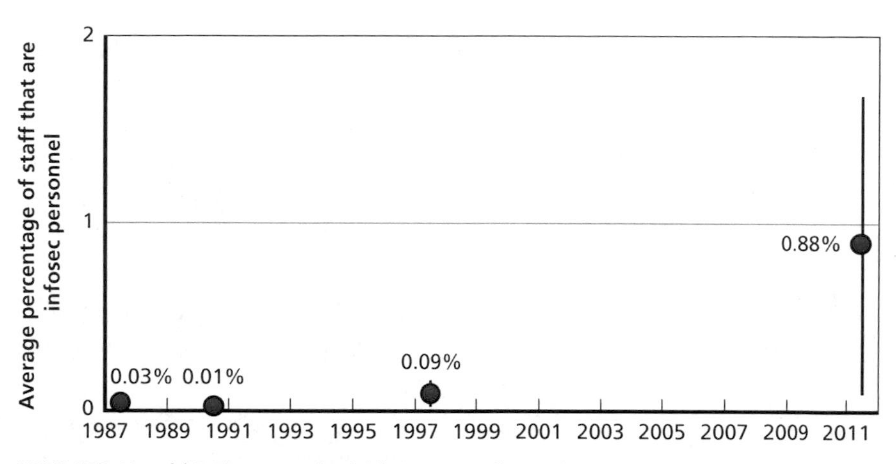

SOURCES: Harold F. Tipton,and Micki Krause, *Information Security Management Handbook*, 6th edition, Boca Raton, Fla.: Auerbach Publications, 2007; Computer Security Institute, *2010/2011 Computer Crime and Security Survey*, New York, 2011; Wood, 2012.
RAND *RR1490-2.5*

human defenders to be constantly active and in constant need of new tools and scripts.[16]

DoD Terms and Definitions

The terms and metrics described in this chapter are more closely associated with private-sector infosec professionals. It is important to note that demands in the private sector affect the supply of personnel for DoD, including the armed services. The main point is that as private-sector demands go up, the supply of talent available to DoD becomes increasingly limited. Nonetheless, it is important to point out specific definitions and job roles that have to be filled by the military, including

[16] Research on the topic of automated exploit generation has identified processes that automate the effort to identify zero-day exploits and harvest these exploitations so that large numbers can be identified. See Thanassis Avgerinos, Sang Kil Cha, Brent Lim Tze Hao, and David Brumley, "AEG: Automatic Exploit Generation," Pittsburgh, Pa.: Carnegie Mellon University, undated.

operations that are more than just the defense-oriented efforts being used in the private sector. We note DoD definitions as follows.

- *Cyberspace operations:* The employment of cyberspace capabilities where the primary purpose is to achieve objectives in or through cyberspace.
- *Cyberspace domain:* A global domain within the information environment consisting of the interdependent networks of IT infrastructures and resident data, including the Internet, telecommunications networks, computer systems, and embedded processors and controllers.
- *Cyberspace workforce:* Personnel who build, secure, operate, defend, and protect DoD and U.S. cyberspace resources, conduct related intelligence activities, enable future operations, and project power in or through cyberspace. It is composed of personnel assigned to the areas of cyberspace effects, cybersecurity, cyberspace IT, and portions of the intelligence workforces.
- *Cybersecurity workforce:* Personnel who secure, defend, and preserve data, networks, net-centric capabilities, and other designated systems by ensuring appropriate security controls and measures are in place, and taking internal defense actions. This includes access to system controls and monitoring, administration, and integration of cybersecurity into all aspects of engineering and acquisition of cyberspace capabilities.
- *Cyberspace effects workforce:* Personnel who plan, support, and execute cyberspace capabilities where the primary purpose is to externally defend or conduct force projection in or through cyberspace.
- *Cyberspace IT workforce:* Personnel who design, build, configure, operate, and maintain IT, networks, and capabilities. This includes actions to prioritize portfolio investments; architect, engineer, acquire, implement, evaluate, and dispose of IT as well as information resource management; and the management, storage, transmission, and display of data and information.
- *Intelligence workforce (cyberspace):* Personnel who collect, process, analyze, and disseminate information from all sources of intelli-

gence on foreign actors' cyber programs, intentions, capabilities, research and development, and operational activities.[17]

Automated Cybersecurity Is the Holy Grail

The private sector is under financial pressure to manage IT costs. Some hope to use automated tools to offset the demand for IT personnel. Figure 2.6 shows the calculated need based on two assumptions: (1) the number of users (e.g., 450,000 AC troops) and (2) the acceptable ratios (e.g., 1:120, 1:80). If any of the trends regarding the needed ratios hold, tens of thousands of infosec personnel will be needed for each component.

Figure 2.6
Information Security Personnel Requirements for 2021 Army End Strength

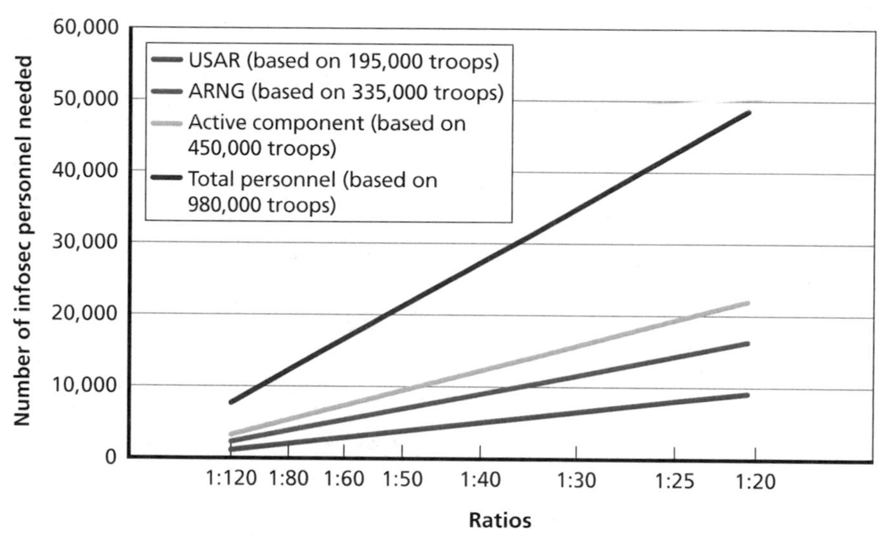

NOTE: Troop figures are 2021 Army end strength estimates (John G. Ferrari, "The Army Program FY16–20," Program Analysis and Evaluation Commission Brief, briefing slides, June 17, 2015, slide 4).
RAND *RR1490-2.6*

---

[17] Definitions adapted from Department of Defense Directive 8140.01, *Cyberspace Workforce Management*, Washington, D.C.: U.S. Department of Defense, August 11, 2015; Joint Publication (JP) 1-02, *DoD Dictionary of Military and Associated Terms*, Washington, D.C.: Joint Chiefs of Staff, February 2017.

The ability to meet such a personnel need is challenging in a fiscally austere budgeting environment.[18] Technological developments might enable some amount of automation that alleviates the demand on infosec personnel. In the interim, however, any unmet need will likely translate into less cybersecurity. More breaches—and perhaps even more mission-ending effects—might occur.

Chapter Summary

Today, the commercial world and government sectors are competing for personnel with cyber skills. We have not assessed how the supply of cyber-skilled personnel will grow in relation to demand; however, it is reasonable to assume from past studies that demand will outpace supply.[19] Online listings for cybersecurity jobs continue to rise.[20] If trends persist, large organizations will need cybersecurity staffs measured in tens of thousands of personnel. Thus far, the federal government has only a fraction of the total workforce it needs, and demand is growing in all sectors. Total demand for personnel with cyber skills could exceed 1 million by 2017.[21] Civilians, contractors, and uniformed personnel will be needed to meet these requirements.

According to LTG Cardon, the Army's needs as of March 2015 were "3,806 military and civilian personnel with core cyber skills."[22] However, given the trends discussed in this chapter, future needs are likely to be higher. Foreign military cyber forces are estimated to be large in number (Table 2.4). While our report does not assess the accuracy of these estimates, these speculations provide some indication of what the United States will be contending with in the future.

---

[18] It may require a restructuring of the forces to meet the cyber personnel requirements.

[19] See Martin C. Libicki, David Senty, and Julia Pollak, *Hackers Wanted: An Examination of the Cybersecurity Labor Market*, Santa Monica, Calif.: RAND Corporation, RR-430, 2014.

[20] U.S. Bureau of Labor Statistics, undated-a.

[21] U.S. Bureau of Labor Statistics, undated-a.

[22] Cardon, 2015. This does not necessarily include thousands of other IT-related skill set holders.

Table 2.4
Open-Source Speculation on Cyber Troops by Country

Country	Speculated Size
Iran[a]	1,500
North Korea[b]	6,000
China[c]	100,000

[a] Yossi Mansharof, *Iran's Cyber War: Hackers in Service of the Regime; IRGC Claims Iran Can Hack Enemy's Advanced Weapons Systems; Iranian Army Official: 'The Cyber Arena Is Actually the Arena of the Hissen Imam,'* Washington, D.C.: The Middle East Media Research Institute, Inquiry and Analysis Series Report No. 1012, August 25, 2013.

[b] Associated Press, "North Korea has 6.000-Strong Cyber Army, says South Korea," January 6, 2015.

[c] John M. McConnell, untitled remarks delivered at the Spring 2015 Senator Christopher S. "Kit" Bond lecture series, YouTube.com, March 12, 2015.

CHAPTER THREE

Findings from the Literature Review

In this chapter, we summarize findings from previous studies and government reports regarding the role of the Army RC in cyber operations and the current state of cyber skills in the RC.

The Reserve Component Will Play an Essential Role in Cyber Operations

In general, RC units are expected to maintain training and qualifications to be available for active duty and/or service to their respective states. With regard to cyber operations, the DoD Cyber Strategy describes the RC as essential in building a cyber workforce prepared to defend the U.S. homeland and U.S. interests from attack in cyberspace.[1] Deliberations pertaining to the RC's role in cyber operations have become increasingly nuanced, and now include exploration of which entities the RC will interact with and what tasks the RC will perform. GEN Keith Alexander, the former commander of U.S. Cyber Command, conveyed this general perspective in testimony to the Senate Armed Services Committee:

> [The reserve components'] role is two-fold . . . one would be how they work with the States, DHS [U.S. Department of Homeland Security], FBI in resiliency, recovery, and help in the investigative portion; . . . and how they work with us in a cyber conflict

[1] DoD, 2015.

> to complement what we're trying to do. We will not have enough force on our side, so we'll depend on the Reserve and National Guard just like the rest of our force structure.[2]

The second question—what tasks the RC will perform—concerns a less well-defined domain. According to U.S. Army Cyber Command and U.S. Cyber Command, RC activities and tasks *mostly* fall under defensive cyberspace operations (DCO). More specifically, the ARNG is expected to provide operational or surge capabilities to others (e.g., U.S. Army Cyber Command)[3] while simultaneously establishing regional cyber capabilities to support the missions of DSCA and homeland defense. Similarly, the USAR is projected to provide increased capacity to conduct the spectrum of cyber missions.

The role of the RC in cyber operations, in addition to being addressed in recent doctrine, is reinforced by multiple studies and commissioned reports. Indeed, a RAND study concluded that "all cyberspace mission areas, especially cyberspace support and cyberspace defense, are suitable for the RC. In fact, elements of the mission are tailor-made for the RC."[4] In short, there is a growing consensus that

---

[2] Keith B. Alexander, "Statement of General Keith B. Alexander, Commander, United States Cyber Command," testimony to the Senate Armed Services Committee, Washington, D.C., March 13, 2013.

[3] The majority of ARNG cyber capability resides in the 54 Computer Network Defense Teams (CND-Ts) with an exclusive focus on DCO and was born from the Y2K requirement. Each state and territory is authorized eight positions per team, but is not required to fill them all. The Virginia Army National Guard Data Processing Unit/Information Operations Support Command conducts full-spectrum cyber operations to support U.S. Cyber Command and others. (Office of the Secretary of Defense, *Cyber Mission Analysis: Mission Analysis for Cyber Operations of Department of Defense*, Washington, D.C., August 21, 2014, p. 3, not available to the general public.)

However, a new construct has been developed in the form of cyber protection teams (CPTs). Ten new National Guard CPTs are expected to stand up in the next few years, and a new CPT has already been stood up in Maryland (the 1636th Cyber Protection Team). Others are slated to be activated in fiscal year 2016 ("Army National Guard Stands Up Cyber Protection Teams, *Army Times*, March 1, 2013).

[4] Al Robbert, James H. Bigelow, John E. Boon, Jr., Lisa M. Harrington, Michael McGee, S. Craig Moore, Daniel M. Norton, and William W. Taylor, *Suitability of Missions for the Air Force Reserve Components*, Santa Monica, Calif.: RAND Corporation, RR-429-AF, 2014.

the RC will play a prominent and essential role in cyber operations in the future. The question of how to sufficiently resource the RC is the next consideration. This question is outside the scope of this report but is discussed in a white paper on this topic:

> To ensure that the RC is properly resourced and trained for Title 10 mission, U.S. Cyber Command and the RC should ensure that cyber units are missioned and resourced using the same process for cyber used for any other mobilization. The cyber units should be identified, have a cyber mission, and a war trace, and be placed in the Army Force Generation (ARFORGEN) model.[5]

Civilian-Acquired Cyber Skills Are Relevant and Valuable to Many Military Cyber Operations

An important consideration in further clarifying the RC's role in cyber operations is understanding the particular skill sets that the RC can contribute. Indeed, there is emerging consensus around the advantages of "citizen-soldiers" who acquire and maintain valuable technical skills in the private sector and apply these skills in their military careers.

In a study conducted for the Air Force, the Institute for Defense Analyses (IDA) specifically examined the extent to which RC personnel can provide unique benefits to the Air Force cyber mission.[6] The study concluded that civilian-acquired cyber skills are relevant

---

[5] Jeff L. Fisher and Brian Wisniewski, *Employment of Reserve Forces in the Army Cyber Structure*, Carlisle Barracks, Pa.: U.S. Army War College, May 2012.

[6] IDA examined the policy question, "To what extent can RC personnel provide unique benefits to the Air Force cyber missions?" Focus groups conducted with 14 subject-matter experts in the Air Force Active, Guard, and Reserve cyber units found that civilian-acquired skills are relevant and valuable across many military cyber operations (e.g., computer network defense exploitation/analysis, computer network attack); however, these skills have less of an impact in others areas (e.g., combat communications). Overall, respondents answered that Air Reserve Component (ARC) civilian cyber skills and operations frequently add value, with 50 percent of subject-matter experts stating "often" (i.e., several times per ARC man-year) and approximately 40 percent stating "constantly" (i.e., every month of ARC member cyber/information operations service). Drew Miller, Daniel B. Levine, and Stanley A. Horowitz, *A New Approach to Force-Mix Analysis: A Case Study Comparing Air Force*

and valuable across many military cyber operations, including computer network defense (CND),[7] computer network attack (CNA),[8] and exploitation/analysis (but not combat communications). Moreover, RC civilian cyber skills and operations frequently add value to the AC. The IDA study reported on focus group results that suggest that civilian skills are applicable in the military context and that these skills are in short supply (Table 3.1).

Table 3.1
IDA Air Force Focus Group Results

Mission	What percentage of ARC cyber warriors have valuable and relevant civilian work skills and experience?	How valuable (on a scale of 0–10) is bringing in these civilian skills and knowledge?[a]
CND	58	8.6
Exploitation/analysis	54	8.3
Network and base operations	47	6.5
Red team inspections	43	6.9
CNA	34	8.1
Information operations (IO)	27	5.1
Combat communications	23	5.1

SOURCE: Miller, Levine, and Horowitz, 2013.

[a] "Since the resource measures for many of the criteria are subjective variables, the alternatives were scored against these criteria on a 0–10 scale of values" (Miller, Levine, and Horowitz, 2013, p. 3).

Active and Reserve Forces Conducting Cyber Missions, Alexandria, Va.: Institute for Defense Analyses, September 2013.

[7] Known as *defensive cyber operations* in Joint Publication 3-12 (R), *Cyberspace Operations*, Washington, D.C.: Joint Chiefs of Staff, February 5, 2013.

[8] Known as *offensive cyber operations* in JP 3-12 (R).

Elements of the Cyberspace Mission Are Tailor-Made for the RC

In addition to technical cyber skills, the RC offers the unique ability to more effectively leverage interpersonal skills to form long-standing relationships with, and deep knowledge of, state and local agencies. This capability is primarily attributed to Army reserve soldiers and guardsmen remaining within a unit for a much longer period than a typical AC rotation.[9]

As noted earlier, the RAND study concluded that

> all cyberspace mission areas, especially cyberspace support and cyberspace defense, are suitable for the RC. In fact, [there are] elements of the mission [that] are tailor-made for the RC: (i) no deployments, (ii) allows the [Air Force] to benefit from developed civilian expertise, (iii) high readiness in most areas due to civilian similarities, (iv) may be appropriate for implementation of sponsored reserve concept, (v) beneficial to state mission and operations.[10]

Training and Maintaining the Skills of IT Personnel Is Difficult and Expensive

Training and maintaining cyber expertise among AC uniformed personnel is difficult, and it requires many resources.[11] For example, training costs are high, in terms of both the time and the expense required to prepare each "cyber soldier." This is true of a number of positions,

---

[9] Craig McKinley, *The National Guard: A Great Value Today and in the Future*, Washington, D.C.: National Guard Bureau, 2011.

[10] Robbert et al., 2014.

[11] Lynn M. Scott, Raymond E. Conley, Richard Mesic, Edward O'Connell, and Darren D. Medlin, *Human Capital Management for the USAF Cyber Force*, Santa Monica, Calif.: RAND Corporation, DB-579-AF, 2012; Fisher and Wisniewski, 2012; James Hosek, Michael G. Mattock, C. Christine Fair, Jennifer Kavanagh, Jennifer Sharp, and Mark E. Totten, *Attracting the Best: How the Military Competes for Information Technology Personnel*, Santa Monica, Calif.: RAND Corporation, MG-108-OSD, 2004; Robbert et al., 2014.

including positions in the Cyber Mission Force and those that support the cyber brigade. It is also a recurring need because the cyber skills of attackers and defenders alike are perishable. The field of IT changes frequently, as do the tactics of attackers and defenders. Thus, technological awareness is vital, and staying "sharp" requires continual training and frequent missions (e.g., "keyboard time").

However, the RC might be at an advantage in this regard, compared with the AC, if—and only if—RC personnel with cyber skills are gaining constant experience and receiving training during their civilian employment. In such cases, the private sector provides employees with valuable training and, in many circumstances, gives them broader exposure to the cybersecurity realm. These individuals are continuously exercising their skills, gaining valuable private-sector experience and then bringing that experience with them when they are exercising (or potentially activated) in a cyber emergency.

The value of experience in cybersecurity is difficult to monetize. It is believed that some ARNG and USAR personnel have gained extensive cyber experience in their public- or private-sector employment outside the military.[12] In some companies, these personnel are seasoned defenders who have likely seen many types of attacks, implemented various lines of defense, and generally observed industry trends over time. So-called amateur hackers (i.e., those not trained to formal standards) likely have useful experience in different areas, such as knowledge of vulnerabilities and usage of specific software packages, even if they are not well versed in more formal concepts of cybersecurity and risk mitigation.[13]

An open question is how easily cyber skills are acquired through training. Many in the cyber community have told us that they believe that cyber skills are best acquired through practice. The thinking is that an individual can learn the fundamentals, but unless he or she is constantly applying that learning on existing systems, the skills will remain undeveloped or erode over time. Members of the Air National

---

[12] Fisher and Wisniewski, 2012.

[13] COL Aida T. Borras, Army action officer for this project, personal correspondence with the authors, November 9, 2015.

Guard refer to this continual honing of skills as "cyber flight hours," analogous to how aircraft pilots gauge experience by tracking flying hours on specific aircraft.

It is worth noting that a cyber range can help with maintaining cyber practice time. The U.S. Army Communications-Electronics Command is making a cyber range available for Army-wide use. The range is intended to provide an "operationally realistic environment with functionality for remote participation."[14] It could provide an opportunity for all soldiers to work on their cyber skills using automated training plans that test skill sets.[15]

Retention Is Difficult in the Face of IT Workforce Demands

Given the permanent-change-of-station rate and the average length of tours, DoD is poised to spend millions of dollars to train AC officer and enlisted personnel in cyber skills—in spite of the certainty that many of them will not remain in the military for an extended period. Some officers and enlisted personnel who received training in the military will remain in an RC capacity. However, retention rates generally hover around 9 percent. Improving retention could result in financial savings and capability gain for DoD, but challenges abound. A significant challenge is money.

Some contend that the private sector pays higher salaries for cybersecurity positions than the uniformed military and the federal government can or choose to offer.[16] Conjecture is that "once active military personnel gain cyber expertise and security clearances, they

---

[14] Douglas A. Solivan, Sr., "Communications-Electronics Cyber Training Range Launches," *Fort Gordon Globe*, July 10, 2015; Borras, 2015.

[15] Borras, 2015.

[16] Dune Lawrence, "The U.S. Government Wants 6,000 New 'Cyberwarriors' by 2016," *Bloomberg Business*, April 15, 2014.

often leave the military for high pay in the private sector."[17] If this is indeed the case, DoD will always find it hard to retain full-time officer and enlisted cyber personnel, and the federal government will struggle to attract good civilian talent. News reports regarding a recent DoD memo highlight current retention issues with regard to highly specialized cyber operators who work for "Red teams" that are reportedly understaffed for the current need.[18]

According to interviews conducted by Hosek et al., the Army and the Air Force have experienced difficulty in retaining IT personnel, in part because

> within both the Army and the Air Force, there was no skill-based special pay for IT or MI [military intelligence] personnel to encourage them to obtain additional training and remain in the military (with the exception of language-related special pay).[19]

A second significant challenge stems from IT workforce demands that are unrelated to compensation. Hosek et al. found that IT personnel

> want access to the latest hardware and software, regular training to keep their skills up to date, flexible schedules that allow them to balance professional life and private life, challenges that keep them motivated (such as a chance to work on hot projects), and the commitment of their employer to help them build an exciting career.[20]

Training and certifying military IT personnel appears, in some cases, to catalyze these trainees' departure from the military:

---

[17] Mikheil Basilaia, *Volunteers and Cyber Security: Options for Georgia*, Tallinn, Estonia: Tallinn University of Technology, 2012.

[18] Shane Harris, "Pentagon Memo: U.S. Weapons Open to Cyberattacks," *The Daily Beast*, December 16, 2015.

[19] Hosek et al., 2004, p. 38.

[20] Hosek et al., 2004, p. 21.

> Once personnel had been trained and gained experience, their private-sector opportunities were tangible and alluring. Skill certification programs and poaching by contractors who provide training were variants on this theme. The military provided skills; skill certification made them more easily transferable to the private sector, and civilian trainers pipelined information about outside pay and job opportunities to trainees.[21]

Chapter Summary and Conclusions

This chapter summarized the findings from a substantial literature review focused on the role of the Army's RC. Congressional testimony and other reports already highlight the essential nature of the RC toward building a cyber workforce. According to the literature, civilian-acquired IT and cyber skills are relevant and valuable across many military cyber operations. The unique ability that the RC offers is related to the long-term assignments that useful reservists provide based on their civilian occupations and/or longer-term assignments when in uniform. Some contend that the private sector pays higher salaries for cybersecurity positions than the uniformed military and the federal government can or chooses to offer.[22] However, salary is not the only significant factor in regard to retention.

---

[21] Hosek et al., 2004, p. 38.

[22] Harris, 2015.

CHAPTER FOUR

Army Reserve Component Cyber Inventory Analysis

Background and Analytical Framework

As threats and opportunities in the cyber domain increase, the Army is working to acquire, train, manage, and develop cyber capabilities across the Total Force. This chapter assessed the quantity (i.e., inventory) and quality of cyber expertise in both the AC and the RC. The first step toward measuring this "inventory" of cyber expertise is identifying personnel with cyber skills that are relevant to the Cyber Mission Force.[1]

Identifying Personnel with Cyber-Related Skills

In the AC, we ascribe cyber skills to an individual who meets at least one of two criteria: (1) is in a cyber-related military occupational specialty (MOS) and (2) has military experience in a cyber-related unit. In the RC, we ascribe cyber skills to an individual who meets at least one of three criteria: (1) is in a cyber-related MOS, (2) has military experience in a cyber-related unit, and (3) has civilian expertise acquired through a cyber-related profession that the individual is currently working in.[2]

[1] Although we examine the concept of "relevant cyber skills" in more depth elsewhere in this report, for the purpose of this chapter, we developed specific definitions of what constitute relevant cyber skills based on the data made available.

[2] We acknowledge that individuals could obtain cyber skills through additional routes or experiences, such as graduation from the Cyber Network Defender course at Fort Gordon.

Having identified three RC avenues for obtaining cyber skills—MOS, unit, and civilian profession—we proceeded to identify the specific MOSs, unit identification codes (UICs), and standard occupational classification (SOC) codes that can be considered cyber-related. Table 4.1 lists cyber-related MOSs by branch.[3] The Army created a similar list as part of its contribution to the "Section 933 report,"[4] which the Office of the Secretary of Defense provided in response to the requirement in Section 933 of the National Defense Authorization Act of 2014 that DoD "conduct a mission analysis of the cyber operations of the Department of Defense."

Table 4.1
Cyber-Related Military Occupational Specialties, by Branch

Signal Corps	Military Intelligence	Functional Areas
255A: Information services technician	352N: Signal intelligence analysis technician	53A: Information systems management
255N: Network management technician	35N: Signal intelligence analyst	24A: Telecom systems analyst
255S: Information protection technician		
25A: Signal officer		
25B: Information technology specialist		

SOURCES: Todd Boudreau, "Cyberspace Defense Technician (MOS 255S)," *Army Communicator*, Vol. 36, No. 1, Spring 2011a; Todd Boudreau, "Cyberspace Network Management Technicians (MOS 255N)," *Army Communicator*, Vol. 36, No. 1, Spring 2011b; Michael Lester, Paul Gross, Carrie McLeish, and Bryan Rude, "Connect: Cyber Support to Joint Information Environment (JIE)," briefing presented at AFCEA TechNet, Augusta, Ga., September 9, 2014, slide 8.

[3] Based on a complete list of cyber-related UICs, we identified 60 ARNG units and 37 USAR units. They include cyber defender, information operations, cyber intelligence, and signal units. Some personnel in these cyber-related UICs are performing functions associated with human resources, public affairs, and supply. However, given that we do not have a precise definition of what is, and is not, a "cyber MOS," we kept those personnel in our data set.

[4] Office of the Secretary of Defense, 2014.

Table 4.2 lists cyber-related SOC codes by major group category.[5] The table comprises all professions that fall under the "Computer and Information Technology" major group category (the SOC 15-1000 series) and includes "Computer and Information Systems Managers" (SOC 11-3021), which falls under the "Management" major group category.

Table 4.2
Cyber-Related Standard Occupational Classifications, by Major Group Category

Computer and Information Technology	Management
15-1111 Computer and information research scientists	11-3021 Computer and information systems managers
15-1121 Computer systems analysts	
15-1122 Information security analysts	
15-1131 Computer programmers	
15-1132 Software developers, applications	
15-1133 Software developers, systems software	
15-1134 Web developers	
15-1141 Database administrators	
15-1142 Network and computer systems administrators	
15-1143 Computer network architects	
15-1151 Computer user support specialists	
15-1152 Computer network support specialists	
15-1199 Computer occupations, all other	

SOURCE: U.S. Bureau of Labor Statistics, "Standard Occupational Classification," web page, undated-b.

[5] Federal agencies use SOC codes, which are managed by the Bureau of Labor Statistics, to classify workers into occupational categories.

Using MOS, UIC, and SOC characteristics, one can identify RC personnel with cyber skills. These individuals will fall into one of three categories: those with purely military cyber skills, those with purely civilian cyber skills, and those with both military and civilian skills (Figure 4.1).

Individuals in all three groups likely have some level of *cyber expertise* (a concept discussed in more detail later). However, individuals with purely military cyber skills are not likely to be using those skills daily when they are not activated. Because cyber skills are highly perishable, these individuals might need additional training to maintain cyber competence.[6] Individuals with purely civilian cyber skills are likely to be highly trained, but their military careers are not aligned with their civilian professions. These individuals could therefore represent untapped potential. Individuals with both military and civilian

Figure 4.1
Categories of RC Personnel with Cyber Skills

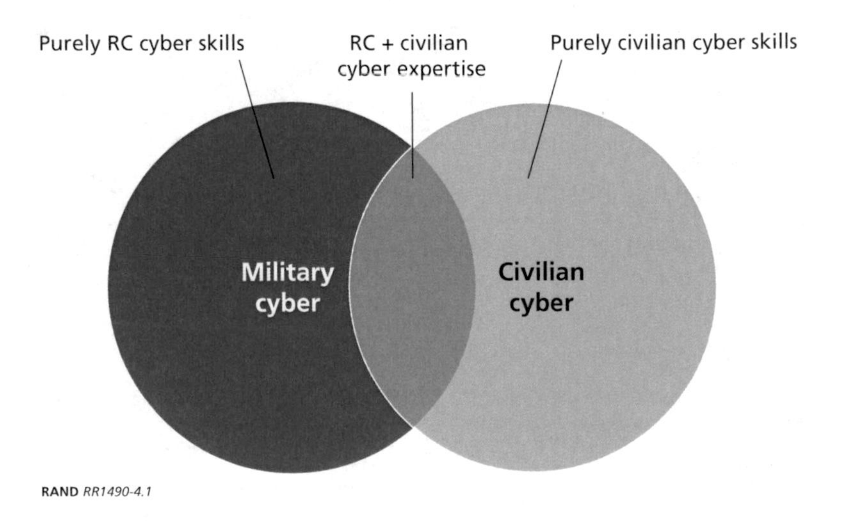

[6] National Research Council, Office of Scientific and Engineering Personnel, *Building a Workforce for the Information Economy*, Washington, D.C.: The National Academies Press, 2001; Timothy R. Homan, "ADP Estimates Companies in U.S. Added 42,000 Jobs," *Bloomberg*, August 4, 2010.

cyber skills exercise those skills in both military and civilian settings. They work on cyber-related issues daily as civilians and can potentially leverage that knowledge while on duty or in training.

In practice, identifying RC personnel with cyber skills is not a completely straightforward process. This is due to gaps in information about RC personnel's civilian professions. Each year, RC personnel must submit information about their civilian profession to the CEI Verification System—the source of much of the data used in our analysis.[7] Once a respondent submits information about his or her civilian profession, using the SOC naming convention and including his or her title and employer, the CEI database autopopulates military characteristics, such as MOS, UIC, service, and component.

In August 2004, DoD established for the Selected Reserve a goal of 95-percent compliance with the requirement to submit civilian employment information into CEI.[8] However, actual response rates for the ARNG and USAR in 2015 were much lower: 55 percent and 15 percent, respectively.[9] Because of these low response rates, our estimates of the inventory of personnel with civilian cyber skills should be interpreted as a lower bound. Additionally, because much of the data are self-reported, there are inherent problems associated with the accuracy of responses. A test-retest study, in which respondents were asked about their occupation several times across a relatively short interval, suggests that this self-reporting error is likely around 25 percent.[10]

---

[7] The CEI database was established under the Uniformed Services Employment and Reemployment Rights Act to help facilitate communication between DoD and the civilian employers of guardsmen and reservists. In particular, the database was established to understand how the activation of personnel might affect their community (e.g., whether activation disproportionately affects first responders in a given area or whether a small company might be unduly burdened).

[8] U.S. Government Accountability Office, *Military Personnel: Additional Actions Needed to Improve Oversight of Reserve Employment Issues*, Washington, D.C., GAO-07-259, February 2007; Pradnya Takalkar, Gordon Waugh, and Theodore Micceri, "A Search for Truth in Student Responses to Selected Survey Items," paper presented at the AIR Forum, Chicago, Ill., May 15–19, 1993.

[9] In 2006, ARNG compliance was actually 93 percent and USAR was 97 percent (U.S. Government Accountability Office, 2007).

[10] Takalkar, Waugh, and Micceri, 1993.

We used the WEX data to backfill our analysis of military-trained cyber personnel. The WEX data are derived from the Defense Manpower Data Center's Active Duty Military Personnel Master File and Reserve Component Common Personnel Data System File. Each record within WEX identifies service, component, reserve category, pay grade, primary service occupation, secondary service occupation, duty service occupation, and UIC. Although the WEX file does not provide information regarding civilian employment, it does track RC personnel who are in a cyber-related MOS or assigned to a cyber-related UIC, as well as the overlap between the two. So, although only approximately 1,000 reservists who submitted their civilian employment information to CEI in the past year are in a cyber-related MOS, WEX indicates that there are actually more than 6,300 reservists currently in a cyber-related MOS. As a result, we use WEX to identify all personnel in cyber-related MOSs and UICs and the CEI database to determine whether those personnel also have civilian cyber expertise. Given the low CEI response rates, this estimate of the inventory of personnel with civilian cyber skills is, again, a lower bound. Later in this chapter, we extrapolate to estimate the number across the entire population of ARNG and USAR.

Description of Data Sets

For ARNG personnel, we had access to two "snapshots" from the CEI database: one from 2011 and one from 2015. The 2015 data set provides information on the current force structure, while comparing the 2011 and 2015 data sets allows for longitudinal analysis of such trends as growth in the population of personnel with civilian cyber skills and the alignment of civilian and military occupations. Table 4.3 compares the characteristics of the 2011 and 2015 ARNG data sets.

For USAR personnel, we had access to a single "snapshot" from the CEI database: one from 2015. The CEI database does not maintain a historical repository of submissions, and we did not have a prior snapshot available. Table 4.4 lists the characteristics of the 2015 USAR data set.

Table 4.3
ARNG Data Sets from the CEI Database

	July 2010–July 2011	December 2013–February 2015
Entries	267,422 (74% of the FY2011 authorized end strength of 361,561)	196,595 (55% of the FY2014 authorized end strength of 354,200)
Respondents in a cyber-related civilian profession	4,554	4,765
Respondents in a cyber-related MOS	4,921 (81% of the total cyber-related MOS force in WEX)[a]	4,256 (65% of the total cyber-related MOS force in WEX)

SOURCE: RAND Arroyo Center analysis of CEI data.

[a] This percentage includes MOS 35F as a cyber-related field; MOS 35F is excluded in other analyses in this table.

Table 4.4
USAR Data Set from the CEI Database

	December 2013–February 2015
Entries	29,625 (15% of the FY2014 authorized end strength of 202,000)
Respondents in a cyber-related civilian profession	1,287
Respondents in a cyber-related MOS	950 (15% of the total cyber-related MOS force in WEX)[a]

SOURCE: RAND Arroyo Center analysis of CEI data.

[a] This percentage includes MOS 35F as a cyber-related field; MOS 35F is excluded in other analyses in this table.

Army Reserve Component Cyber Inventory Analysis, 2015

In this section, we analyze the 2015 CEI snapshots for the ARNG and USAR to estimate the current inventory of personnel with cyber skills. Figure 4.2 shows the civilian professions of the CEI respondents who are in a cyber-related MOS. As one can see, only 35 percent of these guardsmen self-identify as being in a cyber-related civilian profession, compared with 40 percent of reservists.

Of the other 65 percent of ARNG respondents, almost 15 percent are students. Another 10 percent are in the "Does not apply" category, which represents respondents who do not have full- or part-time salaried civilian employment, are not a specified volunteer (i.e., first responder), or do not have student information to report. Given the relatively high rate of "Does not apply" and potential issues associated with self-reported data, we believe that some Active Guard Reserve (AGR) personnel self-identify in this category, despite the fact that AGR personnel are not supposed to submit information into the CEI database. The "Protective services" category captures some cyber-related professions, such as intelligence analysis or criminal analysis, but it also captures AGR personnel who inadvertently put themselves in this category. "Other" contains all other occupations, such as real estate and various blue-collar jobs.

Of the remaining 60 percent of USAR respondents, a much smaller percentage are students or in the "Does not apply" category. Once again, the "Protective services" category captures some cyber-related professions, such as intelligence analysis or criminal analysis.

On average, about 30 percent of ARNG respondents who self-identify as being in a cyber-related profession are also in a cyber-related MOS. Figure 4.3 shows the variation by civilian occupation. These occupations are from the 15-1000 SOC series and the SOC 11-3021 code (Table 4.2). Figure 4.3 also shows how many personnel in each occupation are in a cyber-related MOS (dark green) vs. a non-cyber-related MOS (light green). For infosec analysts, the ratio is 46 percent (143 respondents), which represents a high in these data. The low is only 18 percent for programmers. Those individuals who work in a cyber-related profession but not in a cyber-related MOS should be con-

Figure 4.2
The Civilian Professions of RC Personnel in a Cyber-Related MOS, 2015

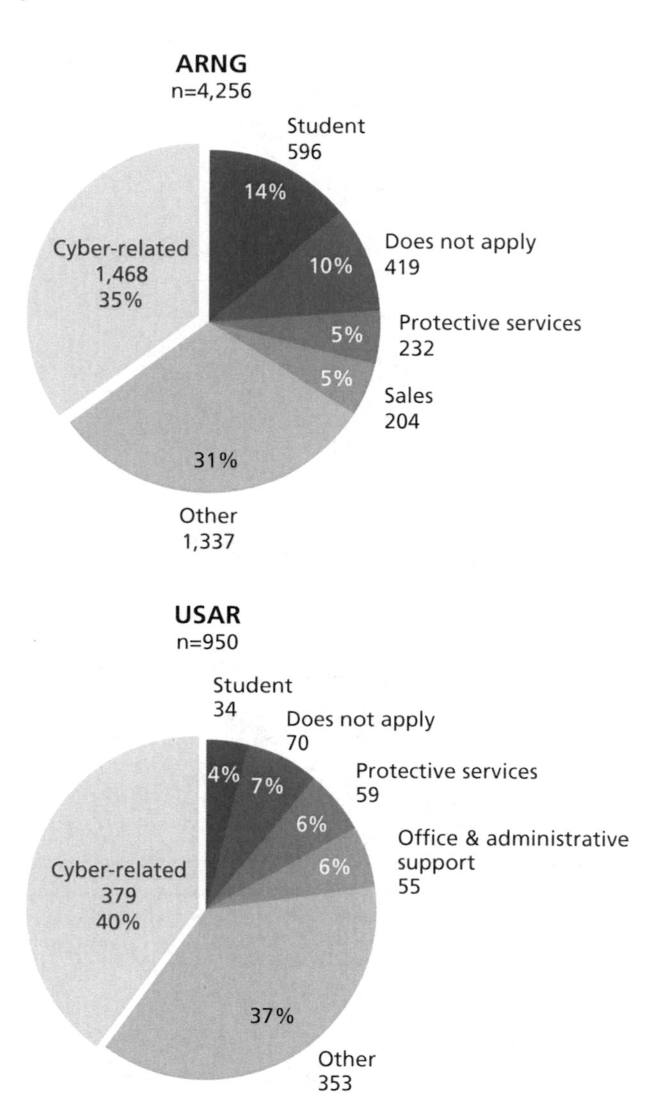

SOURCE: RAND Arroyo Center analysis of CEI data.
RAND *RR1490-4.2*

Figure 4.3
Distribution of ARNG Personnel in a Cyber-Related Civilian Profession, by Occupation and MOS, 2015

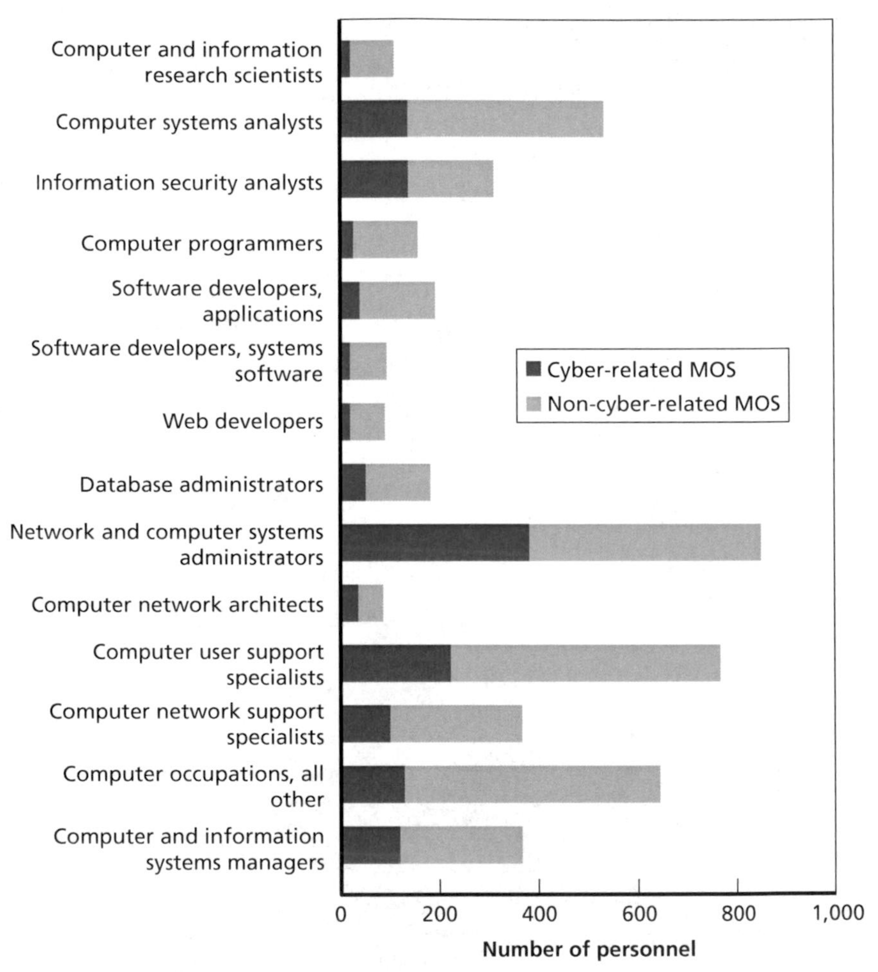

SOURCE: RAND Arroyo Center analysis of CEI data.
RAND *RR1490-4.3*

sidered "untapped potential"—that is, guardsmen whose cyber expertise is not being used by the ARNG.

Figure 4.4 shows the same information for USAR personnel who self-identify as being in a cyber-related civilian profession. As in the case of the ARNG, about 30 percent of USAR personnel who self-

Figure 4.4
Distribution of USAR Personnel in a Cyber-Related Civilian Profession, by Occupation and MOS, 2015

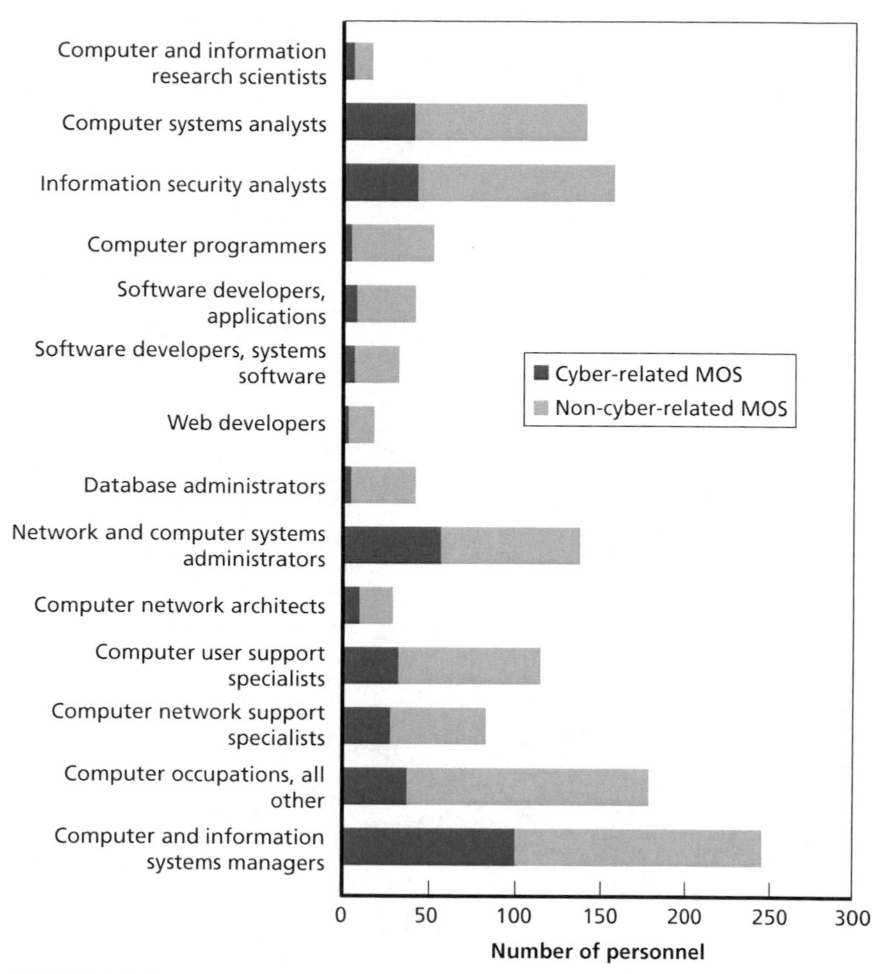

SOURCE: RAND Arroyo Center analysis of CEI data.
RAND *RR1490-4.4*

identify as being in a cyber-related career profession are also in a cyber-related MOS. Again, there is some variation by civilian profession—from 41 percent down to 8 percent.

At this point, we return to the subject of cyber expertise, introduced briefly earlier in this chapter. We propose that RC personnel's degree of cyber expertise can be inferred by looking at the intersections of the three RC avenues for obtaining cyber skills (MOS, unit, and civilian profession). In Figure 4.5, the outer wedges represent personnel who have obtained cyber skills from one avenue only—MOS, UIC, or SOC. While acknowledging the considerable variation in training and experience that these outer wedges comprise, we contend that these personnel have only partial cyber expertise. For example, even a guardsman with ten years' experience as a civilian systems administrator will need training to apply his or her cyber skills to his or her military career. The middle wedges represent personnel who have obtained

Figure 4.5
Assessing Degree of Cyber Expertise

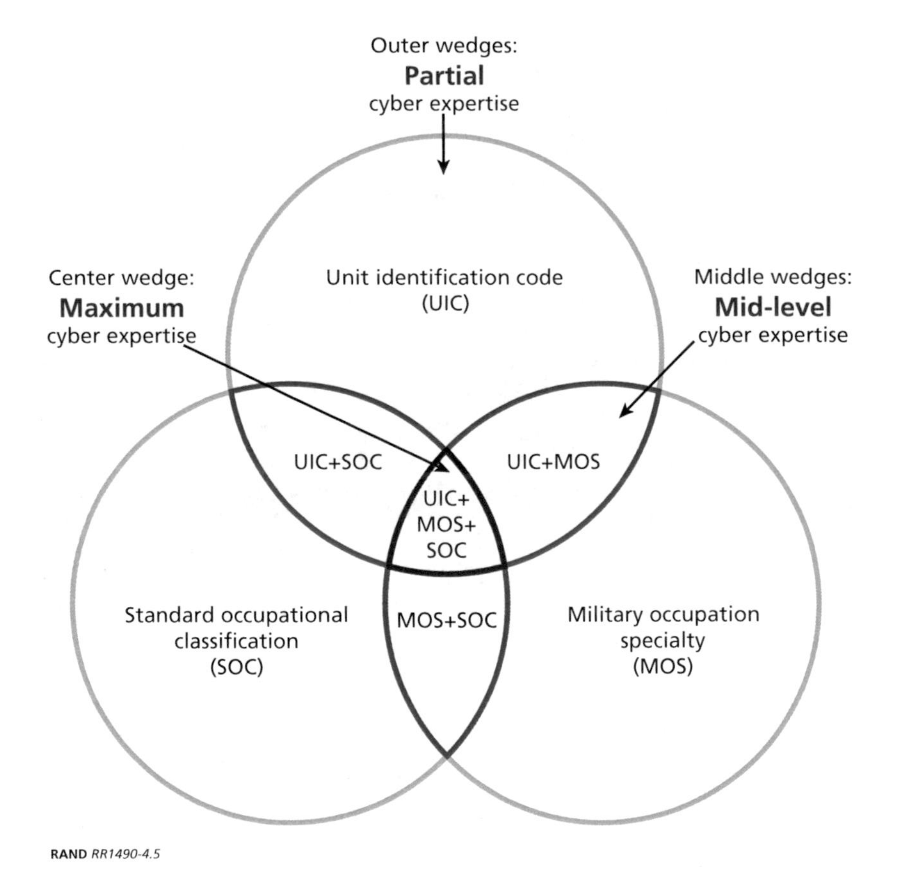

RAND *RR1490-4.5*

cyber skills from two avenues: UIC + MOS, MOS + SOC, or UIC + SOC. We contend that these personnel have mid-level cyber expertise. In the center wedge, where all three avenues intersect, we contend that these personnel have deep or "maximum" cyber expertise.

Figure 4.6 shows the number of ARNG personnel in each of the three expertise wedges, along with the total inventory of personnel with at least partial cyber expertise.[11] As the rightmost column shows, almost 14,700 ARNG personnel have some degree of cyber expertise. Of the 4,765 personnel who self-identify as being in a civilian cyber profession, only 36 percent are also aligned to a cyber-related UIC or

Figure 4.6
Allocation of ARNG Cyber Expertise, by Depth, 2015

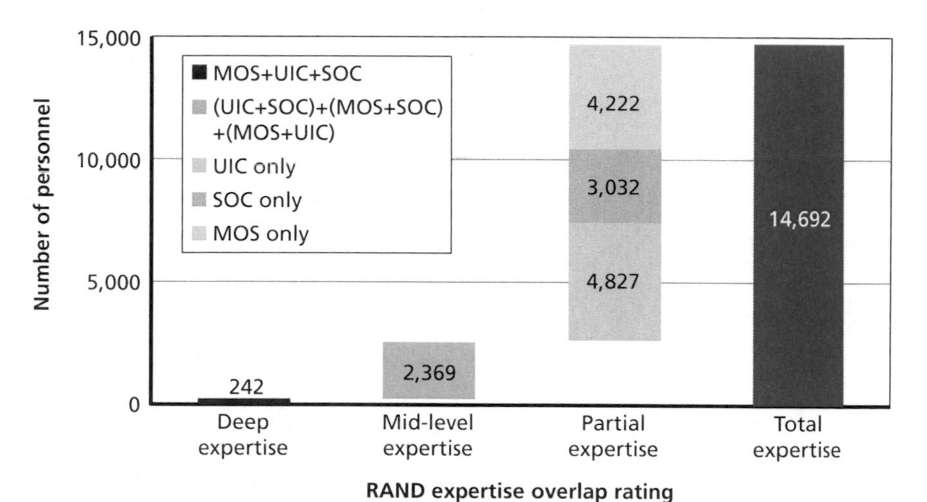

SOURCE: RAND Arroyo Center analysis of WEX and CEI data.
RAND *RR1490-4.6*

---

[11] As noted previously, we used WEX data to partly mitigate the low response rates associated with civilian professional information submitted to the CEI database. Together, CEI and WEX data help flesh out the MOS, UIC, and MOS + UIC wedges for an accurate inventory of the RC force structure when it comes to military cyber-related expertise. However, personnel who fall into a UIC that is cyber-related but not associated with a civilian SOC or a cyber-related MOS may not necessarily be untapped potential. These may represent administrative roles and positions that are not associated with the cyber mission of the unit (e.g., clerks, administrative assistants).

MOS, or both. As the third column shows, 3,032 self-identify as being in a cyber-related civilian profession (the SOC portion of that column), but that skill set is not being leveraged in their military career. The ARNG could leverage the civilian cyber skills of these individuals by aligning them with a cyber-related unit or MOS, or both. This would recategorize more than 20 percent of total ARNG cyber expertise from "partial" to a higher level.

Figure 4.7 shows the same information for the USAR, which has almost 12,000 personnel with some degree of cyber expertise. As in the case of ARNG, only about one-third (35 percent) of USAR respondents who self-identify as being in a cyber-related civilian profession are also aligned to a cyber-related UIC and/or MOS. An opportunity for USAR would be to align the 837 personnel who self-identify as being in a cyber-related civilian profession (but are considered to have only partial cyber expertise) with a cyber-related unit or MOS, or both. This would recategorize a little more than 7 percent of total USAR cyber expertise from "partial" to a higher level.

Figure 4.7
Allocation of USAR Cyber Expertise, by Depth, 2015

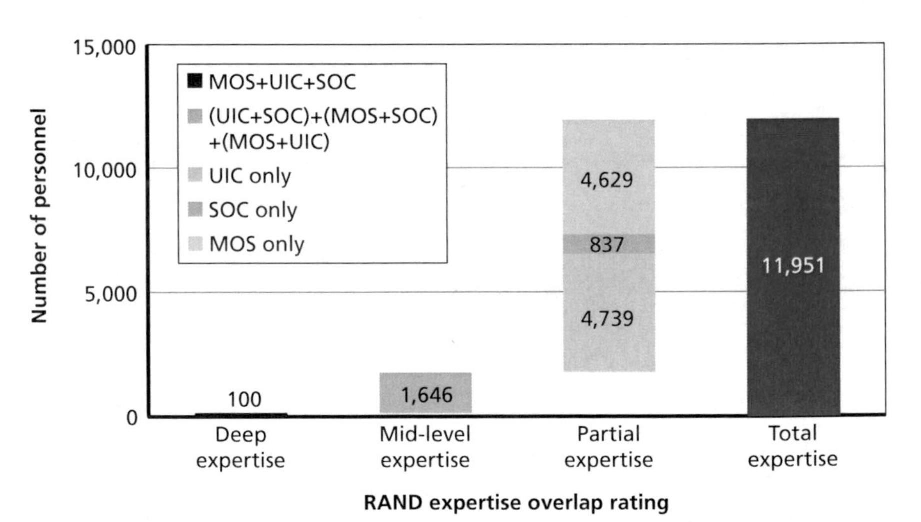

SOURCE: RAND Arroyo Center analysis of WEX and CEI data.

RAND *RR1490-4.7*

Combined, the ARNG and USAR have almost 27,000 personnel with some degree of cyber competence. About 16 percent of those 27,000 personnel have deep or mid-level cyber expertise. The remainder have only partial expertise.

Extrapolating to the Entire RC

The completion rate of the CEI questionnaire was far less than 100 percent. Nonetheless, information gathered from responding soldiers can be used to estimate the distribution of cyber skills in the broader reserve population. For the ARNG, the USAR, and the combined RC, we produced estimates of the percentage of responding soldiers with various levels of cyber skills. We then transformed these estimates into ranges that we have 95 percent confidence hold the true percentage for all RC soldiers. We then applied these ranges to rough assumptions regarding the size of the RC. The volume of responses—more than 200,000 soldiers, or roughly 40 percent of the reserve population—allows us to develop confidence intervals for the estimated ranges, as shown in Table 4.5. From these estimates, we are 95 percent confident that there are roughly 103,000 soldiers with some level of cyber skills in the RC (Table 4.6).

These results are striking. Table 4.7 compares this projected supply with the projected demand for cyber expertise, as discussed in Chapter Two. This projected supply represents potential to fill a future gap for the cyber workforce.

Both the ARNG and the USAR have personnel whose cyber skills could be better leveraged if they were aligned with a cyber-related unit or a cyber-related MOS (or both). This action would mean that more personnel in cyber-related civilian positions would be using their cyber skills in their military careers.

Untapped Potential in the RC

Personnel who use their cyber skills in their civilian occupation (based on their SOC code) and not in their military role represent untapped potential. Based on data in Table 4.5, when we project onto today's RC

Table 4.5
CEI Data Extrapolation to the ARNG and USAR

			Confidence Interval		Population Estimates	
	CEI Number	Proportion	Lower Bound	Upper Bound	Lower Bound	Upper Bound
Army National Guard						
Partial cyber expertise	12,081	0.06	0.06	0.06	20,231	20,942
Mid-level cyber expertise	2,369	0.01	0.01	0.01	3,875	4,198
Deep cyber expertise	242	0.00	0.00	0.00	360	464
Any	14,692	0.07	0.07	0.08	24,646	25,425
N	196,595				350,000	
U.S. Army Reserve						
Partial cyber expertise	10,205	0.34	0.34	0.35	66,117	68,227
Mid-level cyber expertise	1,646	0.06	0.05	0.06	10,326	11,343
Deep cyber expertise	100	0.00	0.00	0.00	529	787
Any	11,951	0.40	0.40	0.41	77,575	79,754
N	29,625				200,000	

SOURCE: RAND Arroyo Center analysis of CEI data.

NOTE: Assumes 350,000 ARNG and 200,000 USAR personnel, for a total RC of 550,000 soldiers. Extrapolations assume a z-distribution to calculate the confidence interval of the proportion of soldiers with cyber skills.

Table 4.6
Sum of Data Extrapolation to Total RC

		Sum of Estimates for ARNG+USAR	
	CEI Number	Lower Bound	Upper Bound
Partial cyber expertise	27,629	86,347	89,169
Mid-level cyber expertise	4,406	14,201	15,541
Deep cyber expertise	259	890	1,251
Any	32,294	102,221	105,179
N	226,220	550,000	

SOURCE: RAND Arroyo Center analysis of CEI data.

NOTE: Assumes 350,000 ARNG and 200,000 USAR personnel, for a total RC of 550,000 soldiers. Extrapolations assume a z-distribution to calculate the confidence interval of the proportion of soldiers with cyber skills.

Table 4.7
Comparison of Projected Supply and Demand

	Expertise Level	Low Estimate	High Estimate	Source
Projected (potential) supply from the RC	Deep	890	1,251	Analysis of CEI data shown in Table 4.5
	Mid-level	14,201	15,541	
	Partial	86,347	89,169	
Potential future demand for the Total Army		49,000		Projection based on vendor surveys and industry trends, as shown in Figure 2.6

SOURCE: RAND Arroyo Center analysis of CEI data.

population, we derive that there are an estimated 10,125 to 11,226 soldiers in the RC that fall into the category of untapped potential. This is shown in Table 4.8.

Army National Guard Cyber Inventory Analysis: Trends

As noted earlier in this chapter, we had access to two "snapshots" from the CEI database for ARNG personnel. This allows us to assess trends in both the growth in the population of personnel with civilian cyber skills and the alignment of civilian and military occupations.

Figure 4.8 depicts the growth of civilian cyber jobs in the United States alongside the growth of ARNG personnel who self-identified as being in a cyber-related civilian profession.[12] As the figure illustrates, the ARNG compound annual growth rate is more than twice that of the nation as a whole. This suggests that ARNG is excelling in capturing the rising tide of cyber capability in the private sector. However, given the previously noted fact that many ARNG personnel in cyber-related civilian positions are not using their cyber skills in their mili-

Table 4.8
Estimated Untapped Potential in the RC

		Population Estimates	
	CEI Number	Lower Bound	Upper Bound
ARNG	3,032	4,984	5,349
USAR	837	6,141	5,877
Total	3,869	11,125	11,226

SOURCE: RAND Arroyo Center analysis of CEI data.

[12] ARNG growth takes into account the different rate of responses between 2011 and 2015. We calculated the rate of response comparison using the cyber-related MOS response rate (81 percent in 2011 and 59 percent in 2015), not the overall ARNG and USAR response rates. We chose this rate based on the assumption that cyber-savvy professionals would have a similar rate of response as cyber-savvy soldiers.

Figure 4.8
Annual Growth Rates in Cyber-Related Civilian Occupations

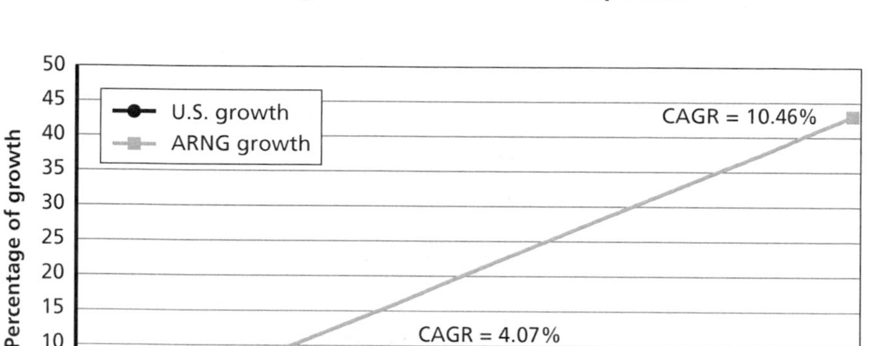

SOURCE: U.S. growth rate data from U.S. Bureau of Labor Statistics, undated-a.
NOTE: CAGR = compound annual growth rate.
RAND *RR1490-4.8*

tary careers, one could infer that ARNG is not adequately leveraging that expertise, once captured.

Clearly, not all ARNG personnel in cyber-related civilian positions are using their cyber skills in their military careers. However, this proportion has been increasing over time. Figure 4.9 shows the self-reported civilian occupations of ARNG personnel who are in a cyber-related MOS. The pie chart on the top illustrates the civilian occupations in 2011; the pie chart on the bottom illustrates the occupations in 2015 (also presented in Figure 4.2). Between 2011 and 2015, the alignment grew from 22 percent to 35 percent.

There has also been an increase in the proportion of ARNG personnel who self-identify as being in a cyber-related profession and are in a cyber-related MOS. Table 4.9 shows, by specific occupation, the number of ARNG respondents who, in 2011 and 2015, self-identified as being in a cyber-related civilian profession. The table also shows the percentage of personnel in each occupation who were also in a cyber-related MOS. The relative populations of the different professions are similar to those seen in 2015; however, the percentage of personnel in a

Figure 4.9
The Civilian Professions of ARNG Personnel in a Cyber-Related MOS, 2011 and 2015

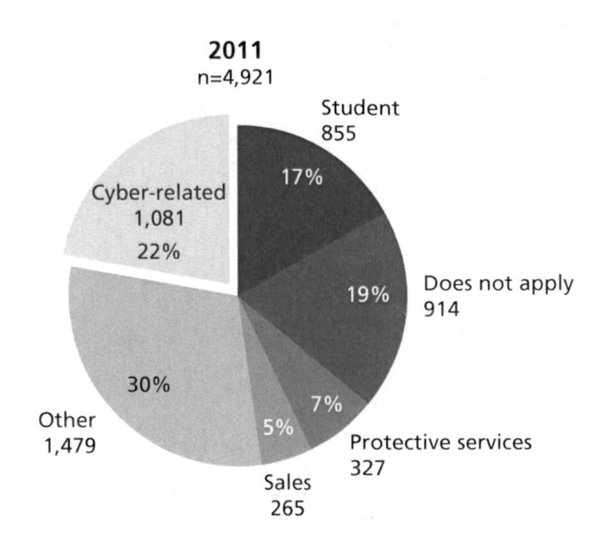

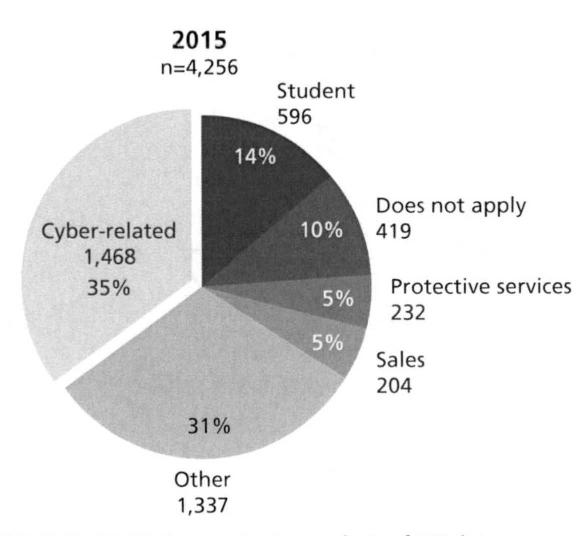

SOURCE: RAND Arroyo Center analysis of CEI data.
RAND *RR1490-4.9*

Table 4.9
Distribution of ARNG Personnel in Cyber-Related Civilian Professions, by Occupation and MOS, 2011 and 2015

	2011		2015	
Civilian Occupation	Total Respondents	% in Cyber-Related MOS	Total Respondents	% in Cyber-Related MOS
Research scientists	135	21	110	21
Systems analysts	589	25	536	26
Information security analysts	275	30	313	46
Programmers	152	14	159	18
Software developers, applications	156	15	196	21
Software developers, systems	68	13	95	22
Web developers	77	19	93	24
Database administrators	196	19	184	29
Systems administrators	795	29	850	45
Network architects	59	24	85	41
User support specialists	543	30	767	30
Network support specialists	358	26	367	28
Occupations, all other	731	18	643	20
Computer and information systems managers	420	21	367	32
Total	4,554	24	4,765	31

SOURCE: RAND Arroyo Center analysis of CEI data.

cyber-related MOS grew between 2011 and 2015. On average, 26 percent of respondents who self-identified as being in a cyber-related profession in 2011 were also in a cyber-related MOS; this figure rose to 33 percent in 2015.

Results for the Active Component

For AC personnel, the depth of cyber skill had to be redefined. We now have only two levels. If an AC soldier is either in a cyber MOS or a cyber unit, that is considered one level (partial). If the soldier is in both a cyber MOS and a cyber unit, that is considered the maximum level. This is shown in Figure 4.10. These markers were chosen based on the databases available.[13] As shown in Table 4.10, there are 37,693 personnel with at least partial expertise.[14]

Chapter Summary and Conclusions

Combined, the ARNG and the USAR have between 101,438 and 105,961 personnel with some degree of cyber competence (see Table 4.6). Both the ARNG and the USAR have personnel with cyber-related skills that could be leveraged if these soldiers were aligned with a cyber-related unit or MOS, or both. A possible opportunity for the RC would be to realign those personnel reporting civilian cyber skills. These actions would mean that more personnel in cyber-related civilian positions would be using their cyber skills in their military careers.

The ARNG appears to be successfully capturing the growth of IT/cyber professionals in the U.S. private sector. However, given the still relatively low number of personnel in cyber-related civilian positions who are using their cyber skills in their military careers, it appears that the ARNG is not completely harnessing those civilian cyber skills to its benefit. On the other hand, data do show improvement between 2011 and 2015 in this regard.

---

[13] There are other characteristics that could imply cyber expertise (e.g., graduates of the Cyber Network Defender course at Fort Gordon).

[14] Of these personnel, 5,061 have maximum expertise.

Figure 4.10
AC Cyber Expertise Is Determined by Degree of Overlap from Training Sources

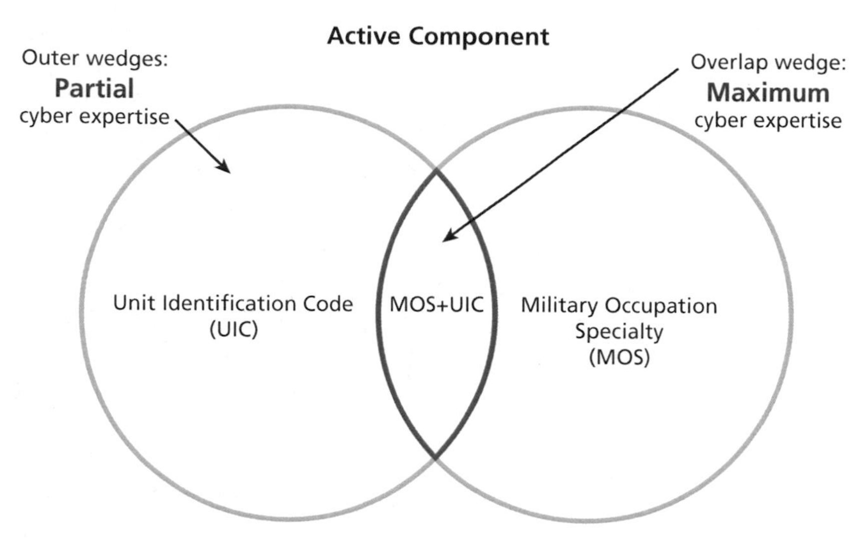

RAND *RR1490-4.10*

Table 4.10
Active Component Personnel with Some Degree of Cyber Expertise

	Number of AC Personnel
In cyber units	13,087
In (potentially) cyber-related MOSs[a]	19,545
Overlapping	5,061
Total	37,693

SOURCE: WEX, June 2014.

[a] This percentage includes MOS 35F as a cyber-related field; MOS 35F is excluded in other analyses in this table.

CHAPTER FIVE

The Role and Importance of Civilian Certification and Training in Developing the Skills Needed for the Cyber Mission Force

In this chapter, we identify the KSAs associated with the Cyber Mission Force, using these KSAs as a surrogate for those that could be associated with cyber operations in general. First, we assess which KSAs are specific to the military. Second, we determine which KSAs could be acquired through civilian certification classes and credentials. Third, we identify the certifications most commonly held by RC personnel by reviewing social media profiles on LinkedIn. Many reservists hold these certifications as part of their civilian occupation, regardless of whether they perform cyber operations in their military duties. From this analysis, we identify the relevant IT and cyber skills that could be gained by "citizen-soldiers" and thus become resident in the entire RC.

Reviewing KSAs Identified for the Cyber Mission Force

To understand the breadth and depth of the cyber-related missions that the U.S. Army might have to conduct, we reviewed U.S. Cyber Command's lists of KSAs associated with the Cyber Mission Force. The

912 KSAs are associated with more than two dozen different personnel roles.[1] Example KSAs include

- knowledge: telecommunication fundamentals
- skills: utilizing enterprise computer forensic tools (e.g., ArcSight, Palantir)
- abilities: conduct vulnerability scans and recognize vulnerabilities in security systems.

Assessing the Military Uniqueness of the KSAs

In reviewing the KSAs, we determined that some are unique to tasks performed by military organizations, some are general-purpose tasks associated with IT jobs in the private sector, and some are very general and not limited to either category.

Results

We found that 21 percent of the KSAs are military, 63 percent are private-sector IT, and 16 percent are general (Figure 5.1). The significance of this finding is that most (about 80 percent) of the criteria associated with cybersecurity job KSAs can be met with content and experience that are available in the private sector.

Examining the Relationship Between Civilian Certifications and KSAs

Background on Certifications

Certifications are often required for infosec-related jobs in the private sector, as well as for a number of cybersecurity jobs across DoD.[2] Many

---

[1] Some KSAs are identical or redundant. Examples of different roles include business data analyst, CND analyst, CND incident responder, cyberspace policy and strategy planner, data administrator, and software engineer.

[2] (ICS)^2 Inc., "DoD Fact Sheet," 2015; Assistant Secretary of Defense for Networks and Information Integration/Department of Defense Chief Information Officer, *Information Assurance Workforce Improvement Program*, Washington, D.C.: U.S. Department of Defense, DoD 8570.01-M, January 24, 2012.

Figure 5.1
Proportion of Cybersecurity Job KSAs, by Category

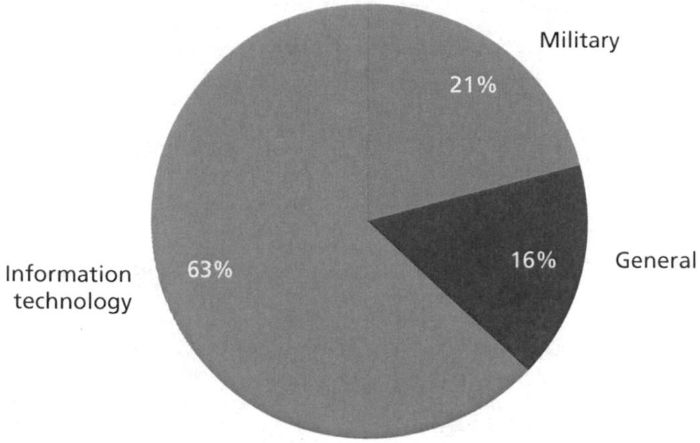

RAND *RR1490-5.1*

of these are listed in Table 5.1. Because there are scores of different certifications, we reviewed a small number of select, relevant certifications and compared them with each KSA in the private-sector IT category, identifying whether each certification covers material related to that KSA. For example, the popular certification Certified Information Systems Security Professional (CISSP) covers 36 percent of the complete set of cyber KSAs (or 57 percent of the private-sector IT KSAs). This leaves 27 percent of the KSAs unaccounted for (see Figure 5.2). The results for other certifications were similar. The significance of this finding is that a civilian with a particular certification would likely be a good candidate—that is, would fulfill a majority of the requirements—for a cybersecurity job in the military.

Our Methodology

We built a spreadsheet that cross-walks select civilian certifications with the KSAs associated with the Cyber Mission Force. We found that most of the key civilian certifications cover most of the 900+ KSAs (Figures 5.3–5.5).

Table 5.1
Certifications

Certification	Topics Covered
CISSP	Security and risk management, asset security, security engineering, communications and network security, identity and access management, security assessment and testing, security operations, software development security
Computer Technology Industry Association (CompTIA) Network+	Installation/configuration of network technologies (including topologies), management, and security
CompTIA Security +	Cloud, Bring Your Own Device, and system control and data acquisition (SCADA) security issues
Certified Ethical Hacker (CEH)	A wide range, including penetration testing, viruses, worms, social engineering, denial of service, and cryptography. For example, • footprinting and reconnaissance • scanning networks; penetration testing • enumeration; system hacking • Trojans and backdoors; viruses and worms • social engineering • denial of service; Structured Query Language (SQL) injection; session hijacking • hacking wireless networks; hacking mobile platforms • evading intrusion detection systems, firewalls, and honeypots • cryptography; buffer overflow

NOTE: *Penetration testing* is the process of attempting to gain access to resources without knowledge of usernames, passwords, and other normal means of access.

Figure 5.2
CISSP Covers 36 Percent of the Total (Cyber) KSAs

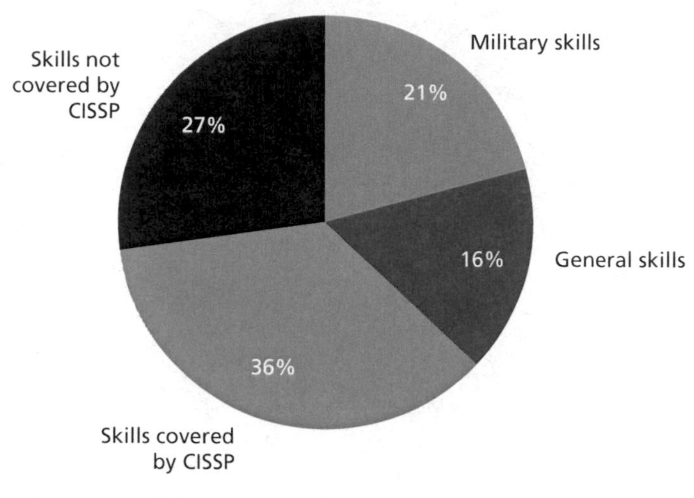

RAND *RR1490-5.2*

Figure 5.3
CompTIA Network+ Covers 27 Percent of the Total KSAs

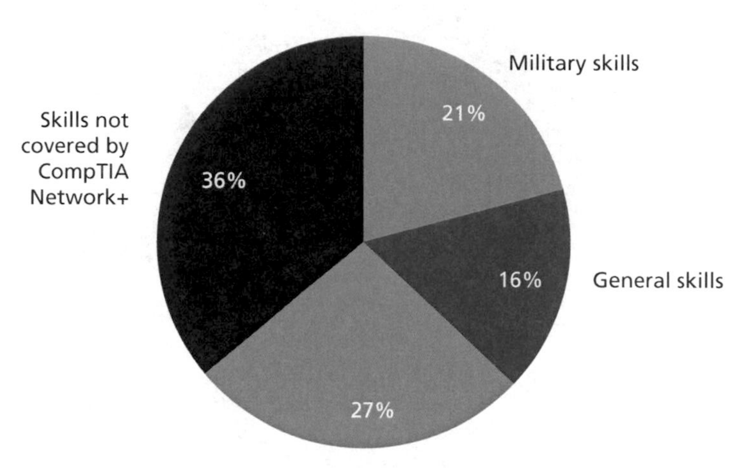

RAND *RR1490-5.3*

Figure 5.4
CompTIA Security+ Covers 42 Percent of the Total KSAs

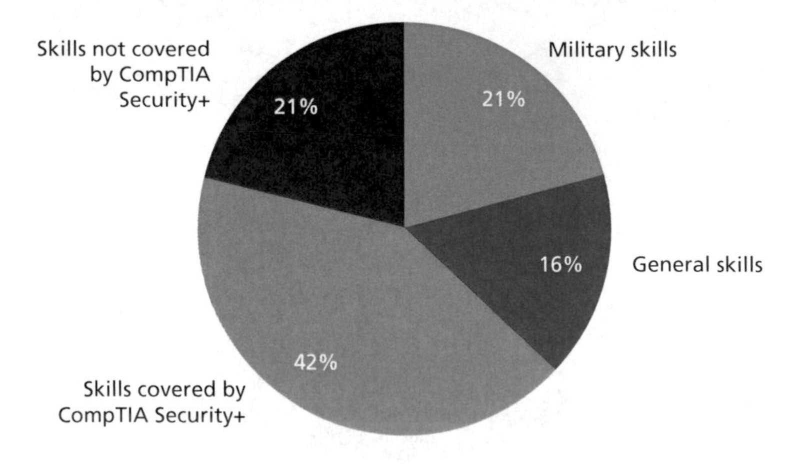

RAND *RR1490-5.4*

Figure 5.5
CEH Covers 47 Percent of the Total KSAs

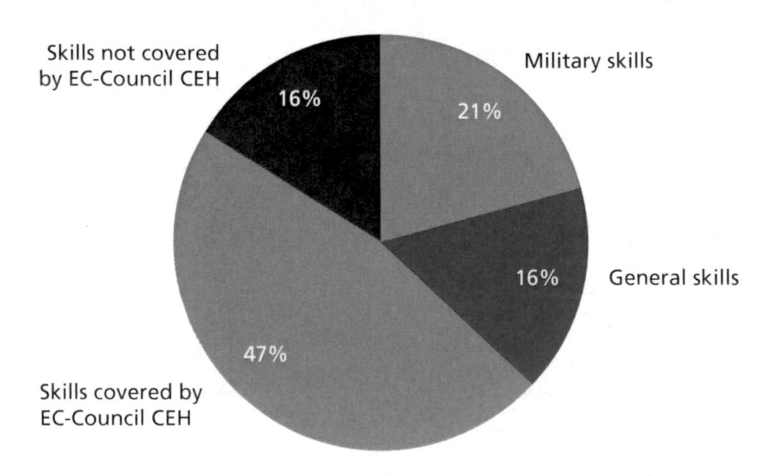

RAND *RR1490-5.5*

Using LinkedIn Data to Identify the Certifications Most Commonly Held by USAR Personnel

To determine which certifications are most common among USAR personnel, we reviewed data from the social-networking site LinkedIn. We used both a macroscopic and microscopic analysis approach. The macroscopic approach involved reviewing a large set (more than 10,000 entries) of LinkedIn profiles using only keyword search techniques.[3]

We performed a scrape of LinkedIn profiles by searching for profiles with "Army reserve" listed as an employer.[4] Table 5.2 shows that a significant portion of the profiles that claim to be those of USAR personnel have cybersecurity certifications (i.e., evidence of cyber skills).

Table 5.3 describes the certifications held by personnel in one of the key USAR cyber organizations: the Army Reserve Cyber Operations Group (ARCOG), which has billets for between 400 and 500 personnel. There is likely overlap between the people who contribute to the data in Table 5.2 and the people who are represented by the data in Table 5.3. But there are clearly more people in the USAR with cyber-related certifications (in Table 5.2) than exist today in the ARCOG.

Chapter Summary and Conclusions

We reviewed and categorized the cybersecurity job KSAs associated with the Cyber Mission Force.[5] In doing so, we learned that most of the KSAs for cyber can be developed outside of military training through civilian experience, training, and education. Our review of common civilian cybersecurity certifications and analysis of their relationship

---

[3] A more microscopic examination of individual LinkedIn profiles is described in Chapter Six. However, that effort was limited to fewer than 2,300 profiles.

[4] Our initial analysis involved searching for Army reserve (specifically "United States Army Reserve") and National Guard, but the numbers we received for the National Guard LinkedIn queries were so small compared with the Army reserve that we decided to focus only on the Army reserve.

[5] Generally speaking, in the Army, MOS qualifications are used to gauge the suitability of individuals to support a mission.

Table 5.2
Scrape of 10,613 LinkedIn Profiles with "Army Reserve" Affiliation

Certification	Absolute Number	Percentage
CompTIA Security+	328	3.1
CompTIA Network+	297	2.8
CompTIA A+	291	2.7
CISSP	109	1
Cisco {Cisco Certified Network Associate [CCNA], Cisco Certified Network Professional [CCNP], Cisco Certified Internet Expert [CCIE]}	{69, 10, 1}	<1
"Microsoft Certified"	60	<1
CEH	33	<1
"VMWare Certified"	7	<1
ISACA Certified Information Systems Auditor (CISA)	3	<1
Systems Security Certified Practitioner (SSCP)	3	<1

SOURCE: RAND Arroyo Center analysis of LinkedIn data.

with the identified KSAs suggests that most of the KSAs that we associate with the general field of IT are indeed taught via popular certifications.[6] Our review of a sample set of USAR personnel with respect to the identified certifications suggests that some USAR personnel (as high as 3 percent for one particular sample of personnel, see Table 5.2) hold relevant cyber certifications.[7]

It must be pointed out that the training associated with work roles for the Cyber Mission Force includes unit and/or group training. This is a very important aspect that is not included in the civilian training opportunities described in this chapter. A theme that we have observed from interviews with government-affiliated Red teams is

[6] Another way of saying this is that advanced cybersecurity certifications (for example, CISSP and CompTIA Security+) cover more than 50 percent of the non-military-specific KSAs.

[7] Of the certification holders, one-half work for the federal government and one-third work in the private sector.

Table 5.3
2014 Report from the Army Reserve Cyber Operations Group

Certification	Absolute Number
CompTIA Security+	157
CompTIA Network+	37
CompTIA A+	19
CISSP	81
CISCO {CCNA, CCNP, CCIE}	{23, 5, not reported}
"Microsoft Certified"	60
CEH	75
"VMWare Certified"	Not reported
ISACA CISA	Not reported
SSCP	Not reported

SOURCE: RAND Arroyo Center analysis of LinkedIn data.

that team dynamics and team roles are vital to the unit's competency. It is most likely that government training utilizes classified materials to provide this level of sophistication, and that this is partly why such training occurs in a classified environment.[8] Therefore, it is not unreasonable to presume that civilian-acquired training cannot fully provide the knowledge and skills needed to counter threats from nation-states.

---

[8] Thomas L. Barnes, information assurance expert at the U.S. Army Cyber Center of Excellence, personal communication with the authors, December 24, 2015.

CHAPTER SIX

Analysis of Reservist Cyber Skills Using LinkedIn Data

In the previous chapter, we broadly analyzed more than 10,000 LinkedIn profiles to determine the type of certifications held by personnel in the RC (e.g., ARNG and USAR personnel). In this chapter, we provide a more in-depth analysis of select LinkedIn profiles in order to assess the specific skills reported by RC personnel. We note that certifications provide an excellent base of knowledge. And, they may be necessary for personnel doing cyber operations. We do not claim that they are sufficient by themselves, but they are, in our opinion, indicators of the potential for further development to perform cyber work functions in support of DoD missions.

Motivation: LinkedIn Offers a Substantial Amount of Relevant Data

According to the U.S. Bureau of Labor Statistics, the 2014 seasonally adjusted average nonfarm payroll employment in the United States was 139 million.[1] According to LinkedIn, there are more than 118 million registered members of the site in the United States.[2] Caution is warranted when comparing these numbers, yet it is clear that the number

[1] U.S. Bureau of Labor Statistics, *Benchmark Information, Comparison of All Employees, Seasonally Adjusted*, 2014a.

[2] LinkedIn, "About LinkedIn," web page, July 2015.

of public profiles on LinkedIn represents a potentially significant source of information.

The number of profiles and their status is constantly in flux, but a snapshot from August 2015 captured 18,410 users claiming current affiliation with the USAR[3] and 30,851 claiming current affiliation with the ARNG.

Table 6.1 summarizes the space of employment, profile availability, and the resultant number of profiles utilized in the analysis. LinkedIn limits the number of profiles accessible through a user's search to approximately 1,000, which explains the numbers in the final two rows of the table. Because of this restriction, our analysis covers only about 3.3 percent of ARNG profiles and 5.4 percent of all USAR profiles, representing less than 1 percent of the total reserve force.

Table 6.1
Summary of Profile Analysis Numbers

Category	Number of People
Average monthly seasonally adjusted nonfarm payroll in the United States in 2014	139 million
Number of LinkedIn profiles in the United States (August 2015)	118 million
Approximate number of USAR profiles on LinkedIn (August 2015)	18,410
Approximate number of ARNG profiles on LinkedIn (August 2015)	30,851
Number of USAR profiles used in the analysis	1,004
Number of ARNG profiles used in the analysis	1,007

SOURCES: U.S. Bureau of Labor Statistics, 2014a; RAND Arroyo Center analysis of LinkedIn data.

NOTE: The analysis considers only those profiles that list either USAR or ARNG as the current employer.

[3] This figure includes both users who claim affiliation with the "United States Army Reserve," which is the official name for the component on LinkedIn, and users who claim affiliation with the "U.S. Army Reserve."

How LinkedIn chooses the 1,000 profiles that appear in a search is not clear, but LinkedIn appears to favor profiles that are in some way connected to the profile of the user conducting the search. Therefore, the results must be considered biased. We also note that our analysis did not distinguish whether each individual profile belongs to that of a uniformed soldier or someone working on a contract for the USAR or ARNG. Finally, it is reasonable to assume that the distribution of experience and experience types among LinkedIn users could be skewed. For example, the level of computer experience of a typical LinkedIn user might be higher or lower than that of the average member of the workforce. Despite these caveats, the data set is still rich.

Results

Our analysis encompasses two areas: skills and employment data. Our skills analysis investigated the self-identified skills within a user's profile. Examples include information assurance and computer security. Our employment data analysis explores such information as where the user is employed, and in what industry. Examples of industries include financial services, defense and space, and law practice.

Skill Analysis

Only some profiles include a list of skills. Of the 2,011 profiles we reviewed (1,004 USAR plus 1,007 ARNG), only 950 listed skill data. Furthermore, skills are not necessarily semantically consistent. That is, although LinkedIn suggests common entries, such as "Security+," users are allowed to enter instead "Security +," "Security Plus," "CompTIA Security+," etc. We controlled for these inconsistencies as much as possible, but the sheer volume of skill data available made this a challenge; for example, the USAR data alone contained 2,648 unique skill entries. We focused on the skills that we deemed most relevant to cyber.

We accounted for all of these factors by establishing a range estimate using (1) the number of profiles that list skills as the denominator for the high value and (2) the total number of profiles as the denominator for the low value (see Table 6.2). By extrapolating these estimates to

Table 6.2
Summary of Skill Analysis

Skill or Certification	Profile Occurrence with Skill Listed (%)	USAR Rank	ARNG Rank
Information assurance	10.3–22.0	1	1
Networking, network administration	7.8–16.4	2	2
System administration	5.1–10.7	6	3
Network security	4.8–10.2	4	5
Computer security	4.8–10.1	3	6
Information security	4.1–8.6	7	8
Security+	3.9–8.3	8	9
Systems engineering	3.8–8.1	5	11
Active directory	3.8–8.0	12	4
IT	3.6–7.6	10	7
Vulnerability assessment	3.0–6.4	9	13
CISSP	3.0–6.3	11	12
Information security management	2.2–4.6	13	15
Cybersecurity, cyber defense, cyber operations, cyber warfare, cyber law, cyber intelligence	1.9–4.1	14	14
Software development, software engineering	1.9–4.1	16	10
Penetration testing	1.0–2.2	15	16

SOURCE: RAND Arroyo Center analysis of LinkedIn data.

NOTE: Where multiple skills are listed, occurrences may not be unique. Red shading indicates skills where the USAR or ARNG ranked lower than the combined rank. Green shading indicates where the USAR or ARNG ranked higher than the combined rank.

the total reserve force, it can be argued that thousands of soldiers have penetration-testing skills or a CISSP certification, or both.

Simple Search

Searching using LinkedIn's simple or advanced search features belies the results of the skills analysis. It does not appear that a profile's "Skills and Expertise" section can be searched using the simple search tool. Doing a keyword search for "penetration testing" on the approximately 50,000 USAR and ARNG profiles yields only 136 matches, or 0.3 percent of the total search. Yet, in our analysis, penetration testing appeared as a skill in 21 out of 950 profiles (2.2 percent). Projecting this result onto the larger population of 50,000 or more profiles translates into hundreds of personnel with this skill. This is a rough approximation.[4]

Employment Data Analysis

The employment analysis investigates the industry and location of each of the 2,011 profiles collected. LinkedIn allows users to select one industry to describe their profile. Table 6.3 highlights the most prevalent industries in the data set.

Over 30 percent of users affiliated with the USAR or ARNG listed military or government administration as their industry. It is difficult to infer much about a person's capabilities based on the industry, because someone could have a technical job in a nontechnical industry. Yet, surprisingly, highly technical skills, such as penetration testing, predominate in technology industries. Eighty-six percent of the profiles that list penetration-testing skills claim to work in the industries of information technology and services and computer and network security (Figure 6.1).

Other more-common skills and certifications, such as vulnerability assessment and CISSP, are more widely distributed in other industries. But, they remain concentrated in the computer and network security and information technology and services industries. Tables 6.4 and 6.5 show the prevalence of these two skills by industry.

---

[4] Spot-checking profiles returned from the keyword search shows that many profiles discuss penetration testing but do not list it explicitly as a skill. It seems logical to conclude that LinkedIn does not include skills and expertise in keyword searches.

Table 6.3
Summary of Employment Industry Analysis

Industry	Occurrence in Profiles (%)	USAR Rank	ARNG Rank
Military	26.3	1	1
Information technology and services	11.0	2	2
Government administration	4.5	5	3
Defense and space	4.4	4	4
Law practice	3.8	3	8
Financial services	2.7	6	5
Human resources	2.2	11	6
Computer and network security	2.1	7	11
Computer software	2.0	9	9
Hospital and health care	1.9	8	15
Logistics and supply chain	1.8	10	14
Management consulting	1.6	12	12
Law enforcement	1.5	15	10

SOURCE: RAND Arroyo Center analysis of LinkedIn data.

NOTE: Red shading indicates skills where the USAR or ARNG ranked lower than the combined rank. Green shading indicates where the USAR or ARNG ranked higher than the combined rank.

Looking at each profile's location affiliation, we see that the Washington, D.C., metropolitan area dominates all of the technical fields (Table 6.6). For the 16 skills identified in this analysis, 20 percent of all profiles surveyed affiliate with the Washington, D.C., metropolitan area.

Figure 6.1
Percentage of Profiles That List Penetration Testing, by Industry

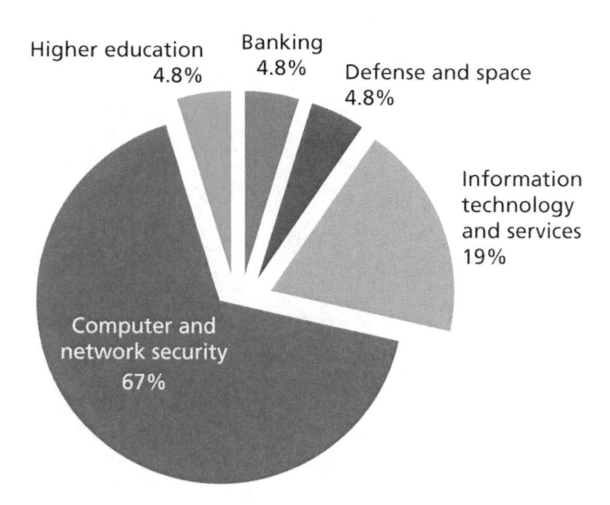

SOURCE: RAND Arroyo Center analysis of LinkedIn data.
RAND *RR1490-6.1*

Chapter Summary

LinkedIn profiles constitute a potentially rich source of information for understanding the cyber skills of the RC. In total, we indexed and analyzed more than 2,000 profiles, representing 4.1 percent of the approximately 50,000 profiles on LinkedIn that identify the USAR or ARNG as a current employer.

Our analysis looked at profile skills, industry identification, and metropolitan location. We focused on 16 technical skills and certifications for the analysis, including information assurance, computer security, CISSP, vulnerability assessment, and penetration testing. Information assurance and networking (or network administration) were the most common skills, at 22 percent and 16 percent, respectively. Penetration-testing skills were evident in over 2 percent of the profiles sampled. Because LinkedIn might attract profiles from technology-oriented individuals, the data might be influenced by selection bias. Nonetheless, when projecting this percentage across the total reserve

Table 6.4
Prevalence of the Vulnerability Assessment Skill, by Industry

Industry	Occurrence in Profiles (%)
Computer and network security	34
Information technology and services	28
Defense and space	8
Military	7
Public safety	5
Financial services	3
Computer software	2
Banking	2
Higher education	2
Management consulting	2
Telecommunications	2
Research	2
Law enforcement	2
Computer networking	2
International trade and development	2

SOURCE: RAND Arroyo Center analysis of LinkedIn data.

force of 550,000 personnel, we are reasonably sure that there are thousands of individuals in the RC with penetration-testing skills.

The most common industries listed for the profiles we analyzed were, in order, military, information technology and services, government administration, and defense and space. Military dominated, at 26 percent. Highly technical skills, such as penetration testing, seemed to be concentrated in technical industries, such as computer and network security and information technology and services. This is consistent with other skills and certifications, such as vulnerability assess-

Table 6.5
Prevalence of CISSP Certification, by Industry

Industry	Occurrence in Profiles (%)
Information technology and services	45
Computer and network security	30
Military	7
Real estate	2
Computer networking	3
Transportation/trucking/railroad	2
Banking	2
Defense and space	2
Telecommunications	2
Higher education	2
Computer software	2

SOURCE: RAND Arroyo Center analysis of LinkedIn data.

ment and CISSP: On average, 75 percent of profiles that list those skills claim affiliation with either the computer and network security or information technology and services industries.

Finally, it is quite clear that the Washington, D.C., metropolitan area is the location that dominates those profiles that list technical skills. In every skill selected, more profiles listed Washington, D.C., than any other area. On average, Washington, D.C., captured 20 percent of profiles per skill. The second-most prevalent areas are Seattle and Raleigh-Durham. In fact, Washington, D.C., is the most prevalent metropolitan area for any skill, with more than twice as many profiles as the next-closest cities (Atlanta and New York). This also might reflect selection bias.

LinkedIn attracts profiles from personnel working in the IT industry and people who work in the Washington, D.C., metropolitan area. It also attracts profiles from personnel affiliated with the military.

Table 6.6
Summary of Skills by Top Three Locations

Skill or Certification	First Location (%)	Second Location (%)	Third Location (%)
Information assurance	Washington, D.C. (23.0)	Five tied (2.4)	—
Networking, network administration	Washington, D.C. (9.7)	Seattle (4.5)	2 tied (3.2)
System administration	Washington, D.C. (10.8)	Seattle (5.9)	Boston (3.9)
Network security	Washington, D.C. (16.5)	Five tied (3.1)	—
Computer security	Washington, D.C. (19.8)	Two tied (4.2)	—
Information security	Washington, D.C. (24.4)	Raleigh-Durham (4.9)	11 tied (2.4)
Security+	Washington, D.C. (20.4)	Two tied (4.3)	—
Systems engineering	Washington, D.C. (30.0)	Denver (3.9)	4 tied (2.6)
Active directory	Washington, D.C. (10.5)	Seattle (6.6)	2 tied (4.0)
IT	Washington, D.C. (23.6)	Pittsburgh (4.2)	11 tied (2.8)
Vulnerability assessment	Washington, D.C. (21.3)	Raleigh-Durham (6.6)	Atlanta (4.9)
CISSP	Washington, D.C. (23.3)	Raleigh-Durham (5.0)	Four tied (3.3)
Information security management	Washington, D.C. (20.5)	Raleigh-Durham (9.1)	Pittsburgh (4.6)
Cybersecurity, cyber defense, cyber operations, cyber warfare, cyber law, cyber intelligence	Washington, D.C. (18.0)	Two tied (10.3)	—
Software development, software engineering	Washington, D.C. (12.8)	Portland (7.7)	8 tied (5.1)
Penetration testing	Washington, D.C. (23.8)	Raleigh-Durham (14.3)	Pittsburgh (9.5)

SOURCE: RAND Arroyo Center analysis of LinkedIn data.

For these reasons and others, it is a useful repository of skills of interest to the Army and analysts studying cyber-related manpower and training issues. We find that the level of cyber expertise that exists in the RC can be estimated with novel uses of social media, such as LinkedIn profiles.

CHAPTER SEVEN

The RAND Arroyo Center Survey of Army Reserve Component Personnel

This chapter reports on a 2015 RAND Arroyo Center survey of RC personnel. The purpose of this survey was to explore current cyber capabilities and untapped cyber potential resident in the RC. Therefore, the survey included a variety of indicators of cyber-related skills and potential to gain cyber-related skills, which included work history, education, self-assessment of cyber-related skills, and life experiences. The motivation for the questions we developed was to get an understanding of the untapped potential that exists in the RC.

Figure 7.1 is a diagram of the sequence of some of the questions in the survey.

Figure 7.1
Types of Cyber-Related Skills

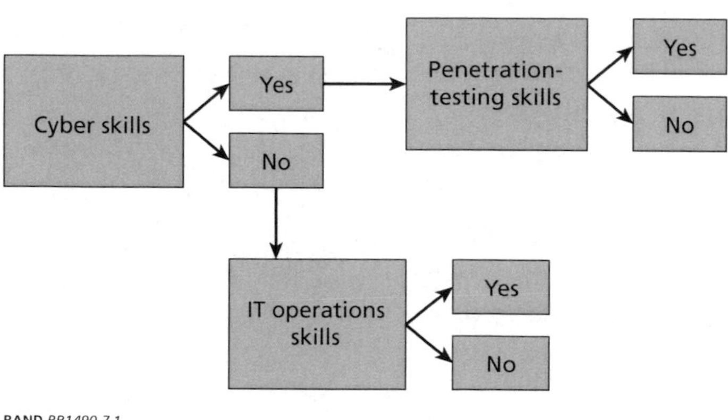

RAND *RR1490-7.1*

Definitions of Cyber-Related Skills Provided to Respondents

Given the goals of the survey (to assess current skills and potential to acquire skills), we comprehensively defined *cyber-related skills* to include cyber skills, penetration-testing skills, and IT operations skills. We provided the following definitions so that survey-takers would have a common frame of reference.

- *Cyber skills:* those skills that directly support defending networks and computers by actively preventing, detecting, identifying, and responding to attacks (impending or in progress). This includes skills that
  - mitigate the impact of an attack (e.g., dynamically reestablish, secure, reroute, reconstitute, or isolate compromised networks to ensure continuous access to the network)
  - are associated with gaining access to networks and other computing devices, including handheld devices for testing (e.g., penetration testing, Red teaming).
- *IT operations skills:* those skills that support operations in cyberspace.
  - Examples include general skills associated with networking and communication, security and compliance, software programming, and website administration.
  - IT operations skills do not include the use of IT to perform ordinary job duties (e.g., sending emails, searching the Internet).

As indicated in Figure 7.1, the survey demonstrated conditional branching depending on the participants' responses. We captured data on whether personnel had cyber skills. We captured data on personnel that held skills that would support operations in cyberspace and/or held skills that would prepare them to acquire cyber skills if they did not already have them. Therefore, if respondents indicated that they did not possess cyber skills, we asked whether they possessed IT operations skills. Additionally, we were interested in identifying those with penetration-testing skills, which we consider to be skills associated with offensive cyber operations. We asked about penetration-testing skills only if participants indicated that they had cyber skills.

Demographics of Respondents

More than 1,200 Army guardsmen and reservists completed the survey. They were predominantly members of the ARNG (over 80 percent), as shown in Figure 7.2.

We invited individual commanders in each state's guard bureau to share the link with that bureau's uniformed personnel in the Army. We also sent emails to the commanding generals of the major RC commands (see Appendix D for the details associated with these invitations). Given how the survey was distributed, through Army officials, we do not know the number of individuals invited to participate in the survey.

Figure 7.2
Number of Respondents, by Component

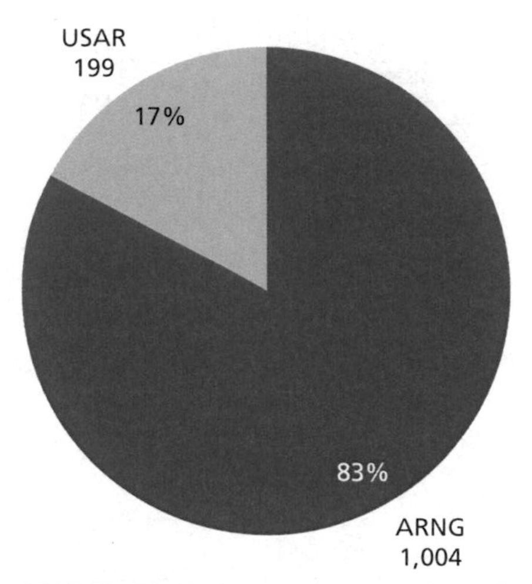

SOURCE: RAND Arroyo Center analysis of survey data.

RAND *RR1490-7.2*

Respondents with Cyber Skills

The fundamental question answered by the respondents was, "Do you have cyber skills?" As Figure 7.3 shows, slightly more than half of the respondents reported having cyber skills. Respondents who said they had no cyber skills were able to claim some level of IT operations proficiency (data not shown).

Respondents were also asked to grade their level of proficiency with regard to their cyber skills. Most of the personnel who reported having cyber skills rated their skills as basic or intermediate, as shown in Figure 7.4.

Respondents were asked about their background with regard to penetration testing, and more than 140 reported having intermediate or advanced skills in this area, as shown in Figure 7.5.

Respondents were asked whether they use their cyber skills only in their Army job, only in their civilian job, only for personal use, or some combination of the three. As Figure 7.6 shows, a large portion of

Figure 7.3
Percentage of Respondents With and Without Cyber Skills

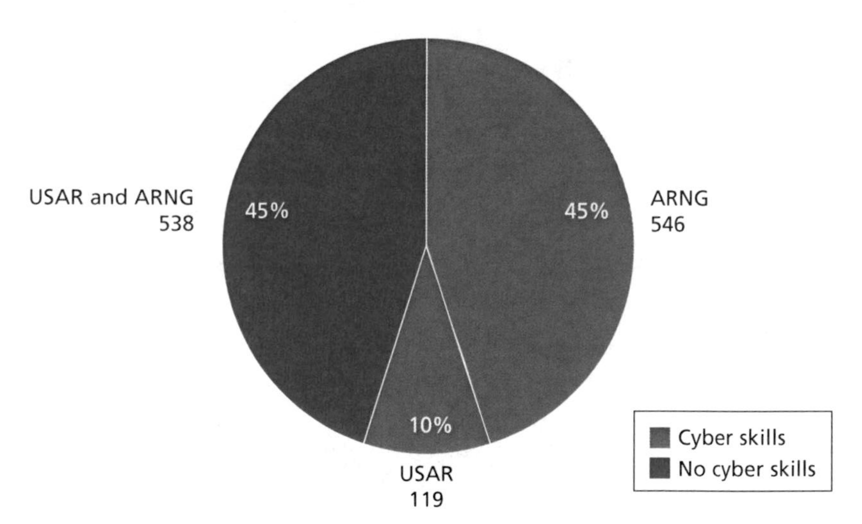

SOURCE: RAND Arroyo Center analysis of survey data.
RAND *RR1490-7.3*

Figure 7.4
Number of Respondents, by Cyber Skill Level and Component

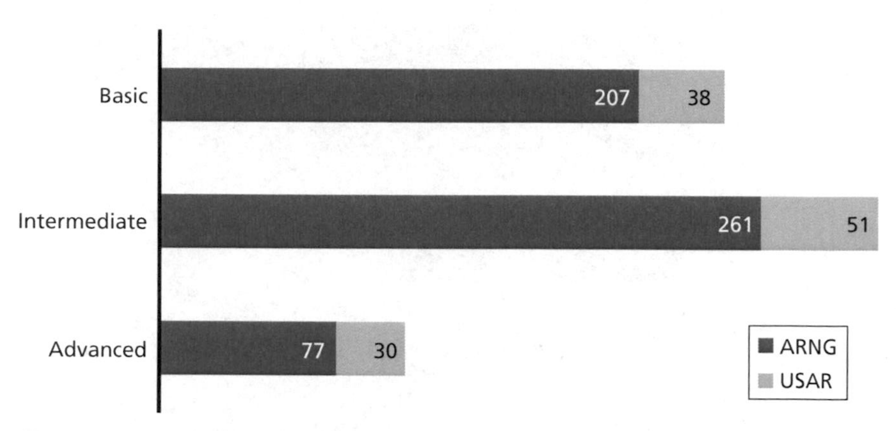

SOURCE: RAND Arroyo Center analysis of survey data.
RAND *RR1490-7.4*

Figure 7.5
Number of Respondents, by Penetration-Testing Skill Level and by Component

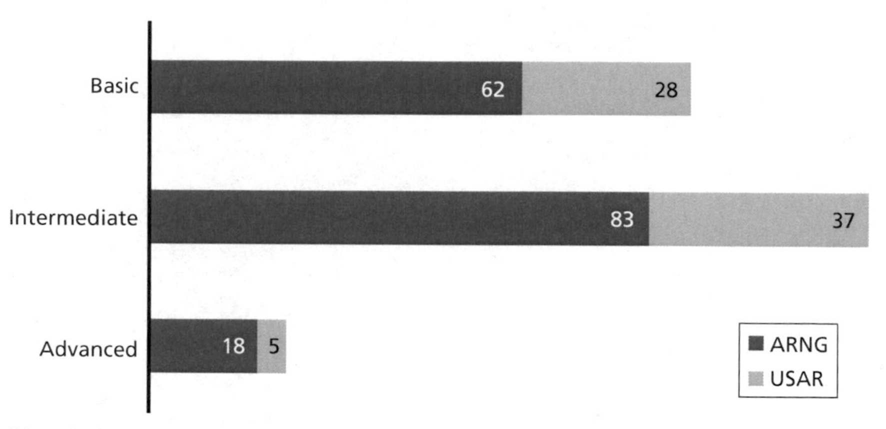

SOURCE: RAND Arroyo Center analysis of survey data.
NOTE: There were 235 total respondents who reported having penetration-testing skills, but only 233 indicated their level of skill.
RAND *RR1490-7.5*

Figure 7.6
Number of Respondents, by Cyber Skill Use and Job Category

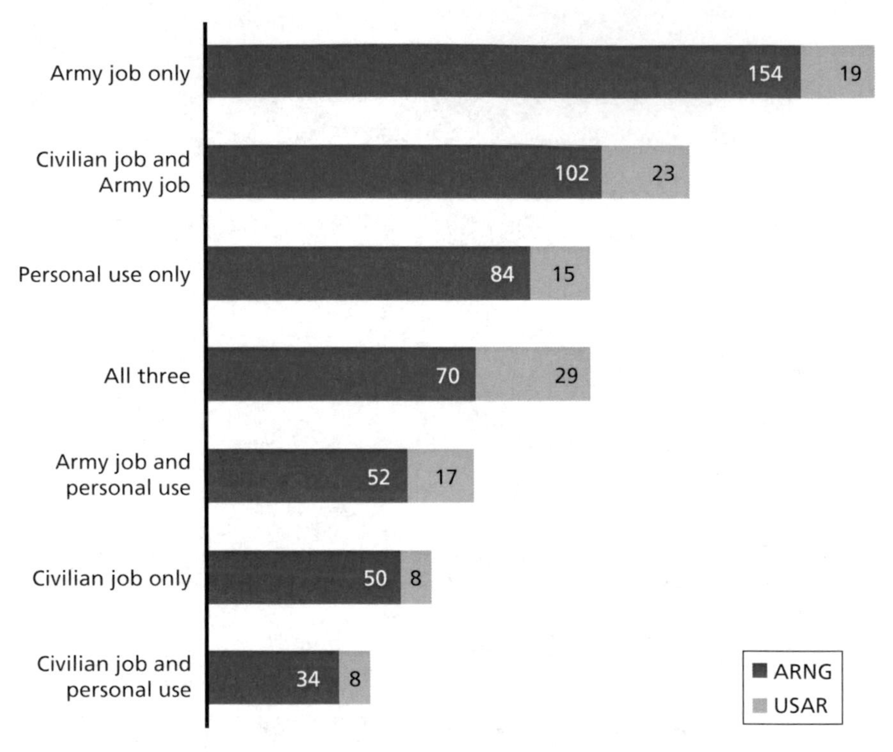

SOURCE: RAND Arroyo Center analysis of survey data.
RAND *RR1490-7.6*

respondents said that they use their skills only in their Army job and not in their civilian job.

Respondents were also asked whether they use their penetration-testing skills only in their Army job, only in their civilian job, or in both jobs. Figure 7.7 shows that there are appreciable numbers of personnel who use penetration-testing skills in their civilian job but not in their Army job.

Among respondents who do not use their cyber skills in the Army, the most commonly cited reason is the lack of a cyber job in their unit (Figure 7.8). It is also interesting that few say that they do not want to use their cyber skills in their Army job. To reinforce this observation,

Figure 7.7
Number of Respondents, by Penetration-Testing Skill Use and Job Category

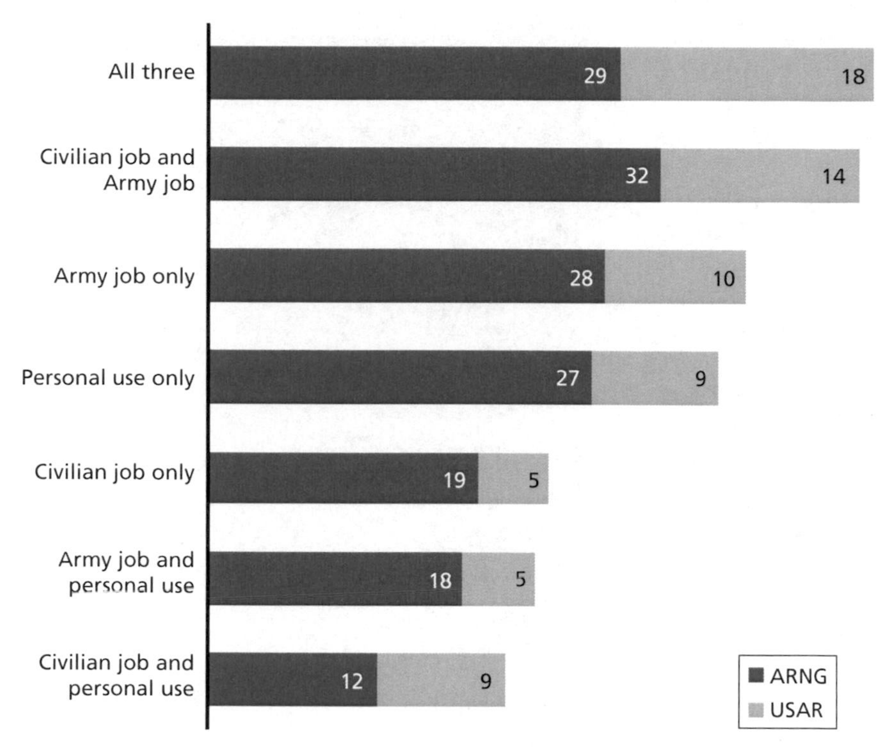

SOURCE: RAND Arroyo Center analysis of survey data.
RAND *RR1490-7.7*

Figure 7.9 shows the majority of respondents in this group are interested in using their cyber skills in the Army.

Many respondents who indicated being only moderately interested or not at all interested in using their cyber skills in the Army reported that an incentive would encourage them to seek this option (Figure 7.10). The most popular incentives were additional training, financial compensation, and promotion and advancement opportunities.[1]

[1] Respondents were allowed to choose more than one incentive.

Figure 7.8
Number of Respondents, by Reason for Not Using Cyber Skills in the Army and by Component

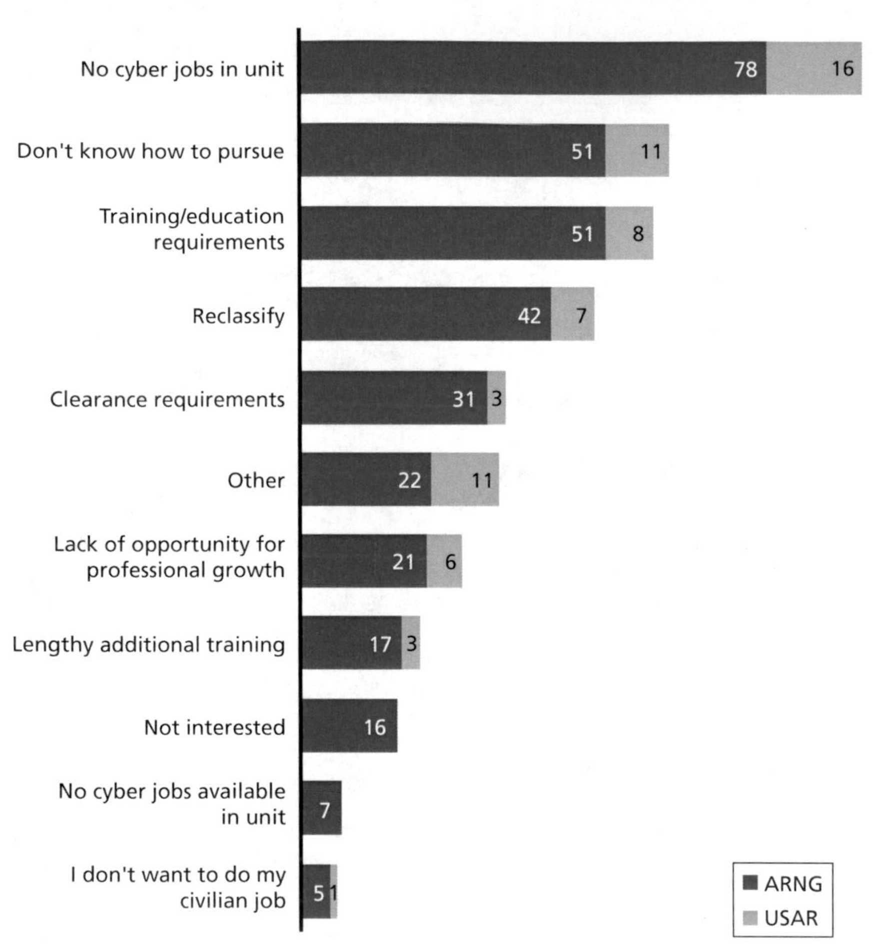

SOURCE: RAND Arroyo Center analysis of survey data.
RAND *RR1490-7.8*

Figure 7.9
Number of Respondents Who Are Not Currently Using Their Cyber Skills in the Army but Report Interest in Incentives to Do So

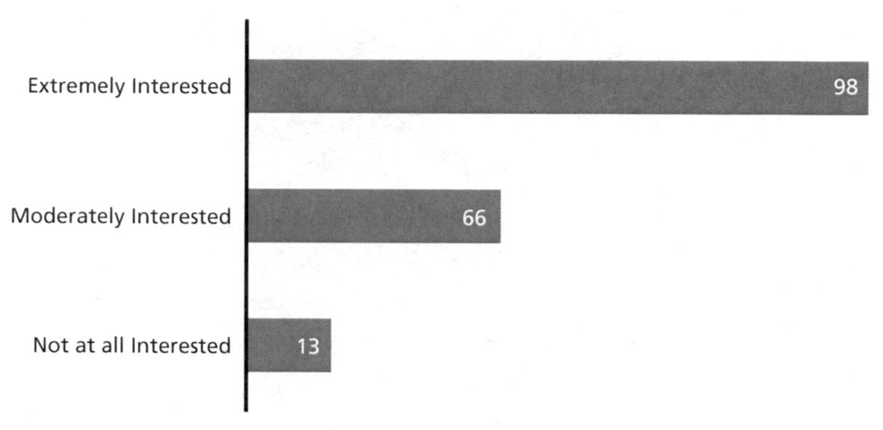

SOURCE: RAND Arroyo Center analysis of survey data.
RAND *RR1490-7.9*

As we note in earlier chapters when analyzing LinkedIn, and as previous surveys (e.g., ARCOG briefings) have indicated, a number of Army guardsmen and reservists hold Security+, Network+, and CEH certifications (Figure 7.11).

Respondents with intermediate or advanced cyber skills were asked to report their education level (Figure 7.12). The responses range in general from high school diplomas to doctorate degrees.

As Figure 7.13 shows, most of the respondents with intermediate or advanced cyber skills said that they are in civilian occupations related to computers or mathematics.[2]

Figure 7.14 describes the number of individuals who responded to our survey and indicated that they held multiple certificates. Based on the data in the chart, approximately 16 percent of the respondents hold multiple certifications.

---

[2] The computer and mathematical–related occupation category includes computer and information research scientists, software developers and programmers, database and system administrators, actuaries, mathematicians, operations research analysts, and statisticians.

Figure 7.10
Number of Respondents with Moderate or No Interest in Using Their Cyber Skills in the Army but Interest in Incentives to Do So, by Component

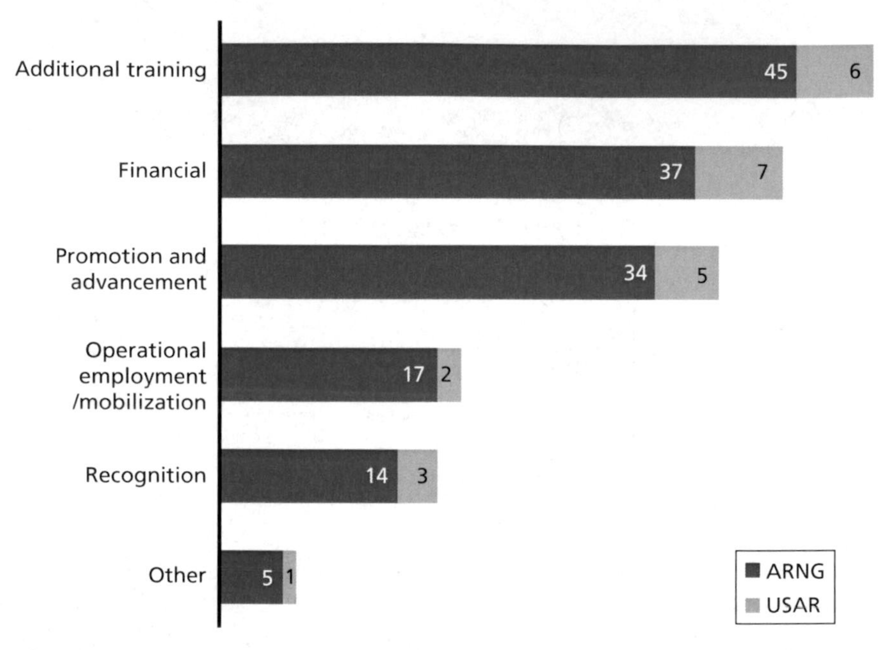

SOURCE: RAND Arroyo Center analysis of survey data.
RAND *RR1490-7.10*

Figure 7.11
Number of Respondents Who Hold a Certification, by Certification

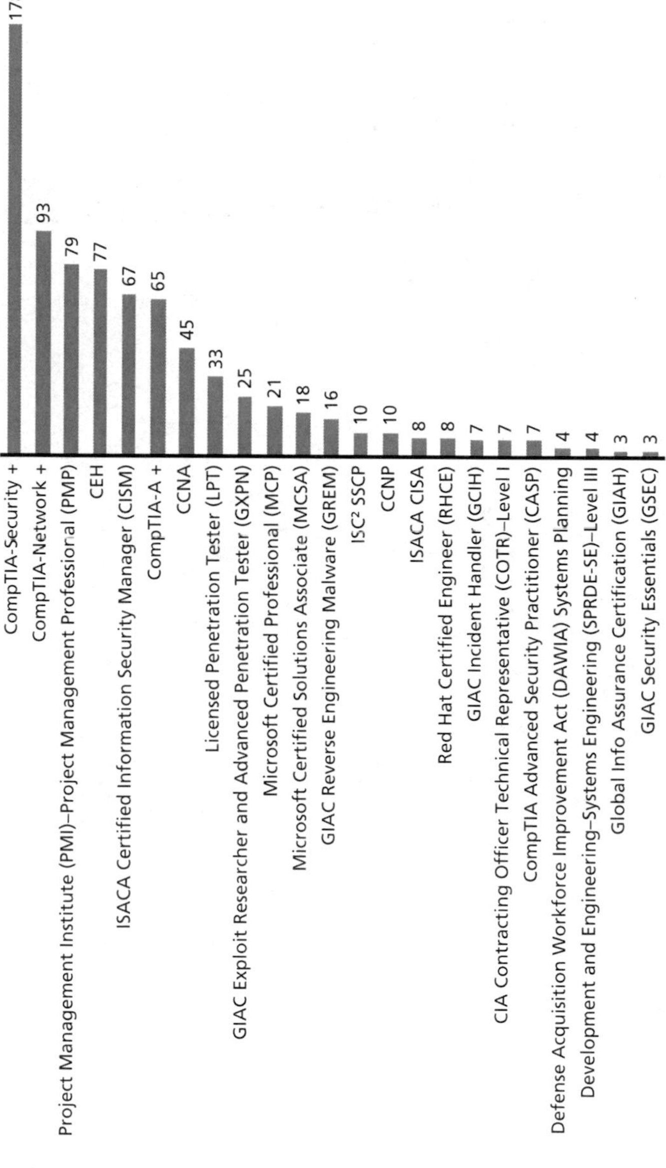

SOURCE: RAND Arroyo Center analysis of survey data.
NOTE: GIAC = Global Information Assurance Certification.
RAND *RR1490-7.11*

Figure 7.12
Number of Respondents with Intermediate or Advanced Cyber Skills, by Education Level

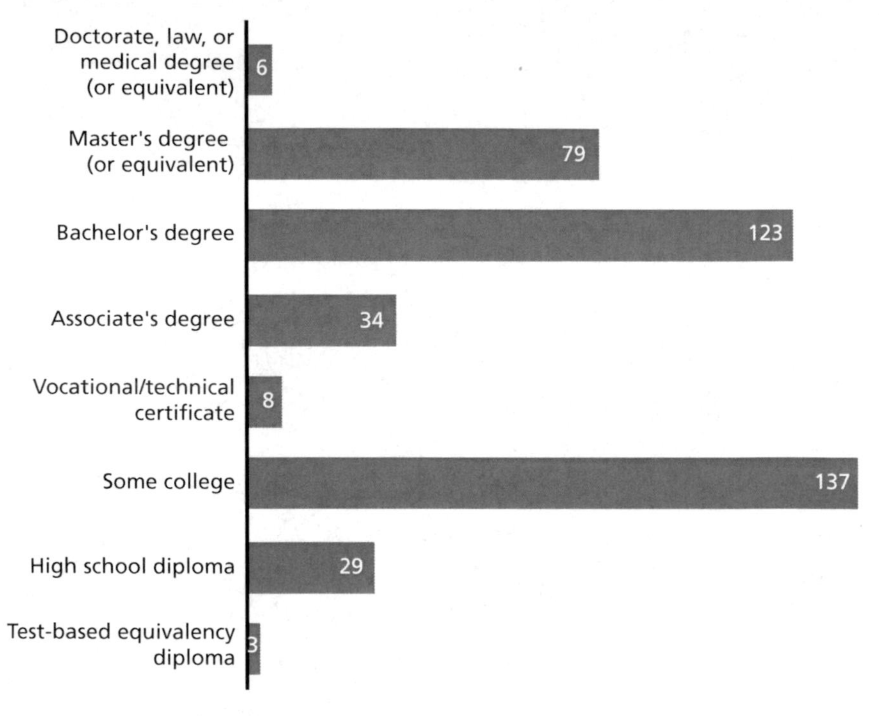

SOURCE: RAND Arroyo Center analysis of survey data.
RAND *RR1490-7.12*

Figure 7.13
Number of Respondents with Intermediate or Advanced Cyber Skills, by Occupation

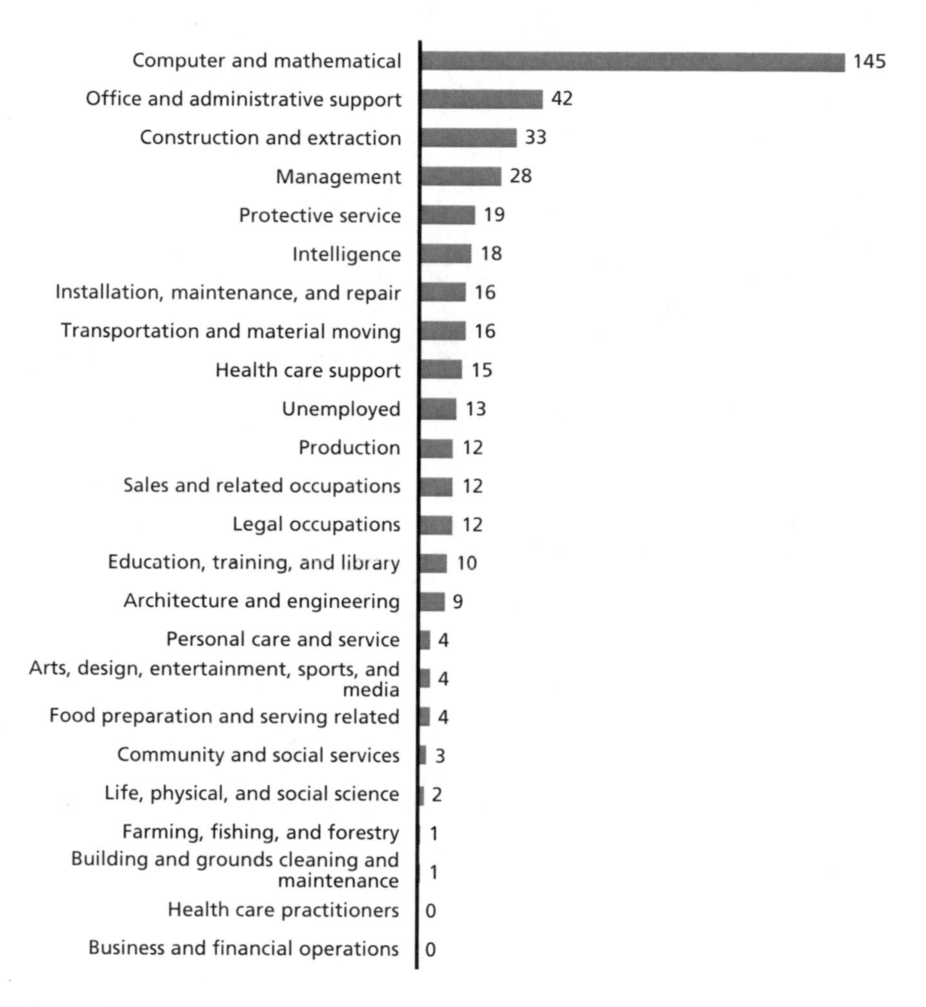

SOURCE: RAND Arroyo Center analysis of survey data.
RAND *RR1490-7.13*

Figure 7.14
Percentage of Respondents Who Hold Multiple Relevant Certifications

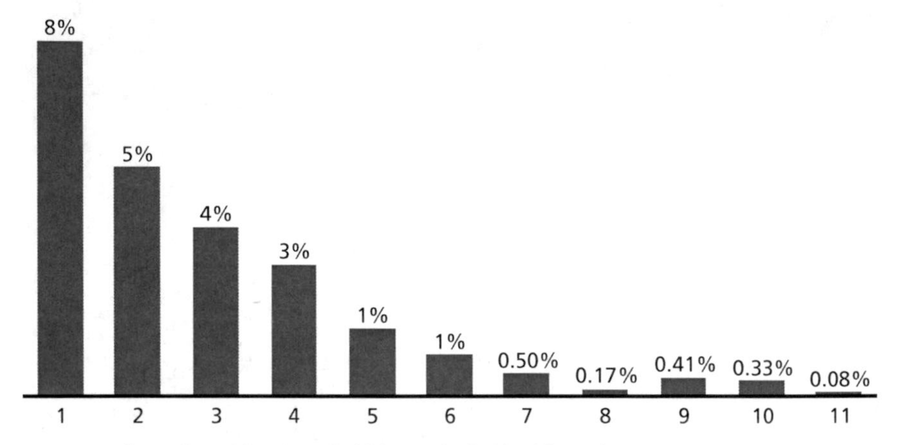

SOURCE: RAND Arroyo Center analysis of survey data.
RAND *RR1490-7.14*

Chapter Summary and Recommendations

Summary

In 2015, more than 1,200 Army guardsmen and reservists participated in a brief survey designed to explore which cyber-related skills are resident in the RC and how much potential is untapped. More than 400 respondents reported having intermediate or advanced cyber-related skills, and 235 self-identified as possessing penetration-testing skills. Although the majority of respondents (70 percent) said that they use their cyber-related skills in the Army, some do not, indicating that those skills are unutilized.

Importantly, a number of respondents (n = 99) reported using their cyber-related skills in only the "other" domain (e.g., open-source code, personal interests), a use that is not documented in Army records (both military and civilian). They too represent untapped potential.

Respondents reported being overwhelmingly interested in applying their cyber-related skills in the Army, but the lack of cyber jobs in their unit is cited as the primary barrier. A substantial number of respondents also reported confusion about how to pursue a cyber career in the Army.

Recommendations

There is value in a detailed survey of cyber professionals that examines specific training, education, and experiences. The Army should conduct surveys of this depth on a more regular basis, although it does not seem practical to us to expand the mandatory CEI questionnaire to this level of detail, given the need to keep such questionnaires short. The RC should seek to perform a survey like the one described in this chapter that reaches the majority of members of the Army Reserve and National Guard. It should include training, education, and experience standards.

Future Work

Survey participants were asked to report cyber-relevant life experiences (i.e., biographical data) by selecting from a set of nearly three dozen possible options, including the following questions:

- Have you ever used networking tools (e.g., Kismet, Snort, tcpdump) in the past?
- Have you ever used scripting languages (e.g., shell, bash, PERL, Python)?
- Have you ever used the "command line" in the past?

In addition, participants were allowed to indicate their level of experience (e.g., "I have done once or twice," "I have done three times or more") for each of the questions answered in the affirmative. This report does not provide any analysis from responses to these questions. However, RAND Arroyo Center will examine the data in depth in future work. This will include a comparison of the experiences of respondents for varying cyber skill levels. Our approach may prove to be a promising one for future identification of cyber-related potential and yield insights regarding ideal candidates for the Cyber Mission Force or other Army cyber personnel.

Figure 7.15 is an example of data we collected from the respondents. The red bars represent the experiences of the respondents with advanced or intermediate cyber skills. The blue bars represent the experiences of the respondents who reported having "no cyber skills." Clearly, there are a set of experiences that are more common to respondents with advanced or intermediate cyber skills. This is an example of how the data we are collecting can be used to gain insights.[3] Note that not all of the biodata items we collected are shown in Figure 7.15: The ones shown are the ones that we can report on, and those not shown are not available for public disclosure.[4]

[3] According to Reed, the government of Israel uses such an approach to identify cyber talent for one of its key cyber agencies: "Applicants are admitted only after an online questionnaire, followed by a battery of more rigorous tests to gauge their abilities in programming, languages and thinking outside the box" (John Reed, "Unit 8200: Israel's Cyber Spy Agency," *Financial Times*, July 10, 2015).

[4] Other biodata items we collected are used in a U.S. Air Force survey effort and cannot be disclosed.

Our approach was motivated in part by the work of Trippe et al., which addressed an Armed Services Vocational Aptitude Battery (ASVAB) Review Panel recommendation to develop better tests of cyber skills (D. Matthew Trippe, Karen O. Moriarty, Teresa L. Russell, Thomas R. Caretta, and Adam S. Beatty, "Development of a Cyber/Information Technology Knowledge Test of Military Enlisted Technical Training Qualification," *Military Psychology*, Vol. 26, No. 3, 2014).

Figure 7.15
Sample Biodata Items

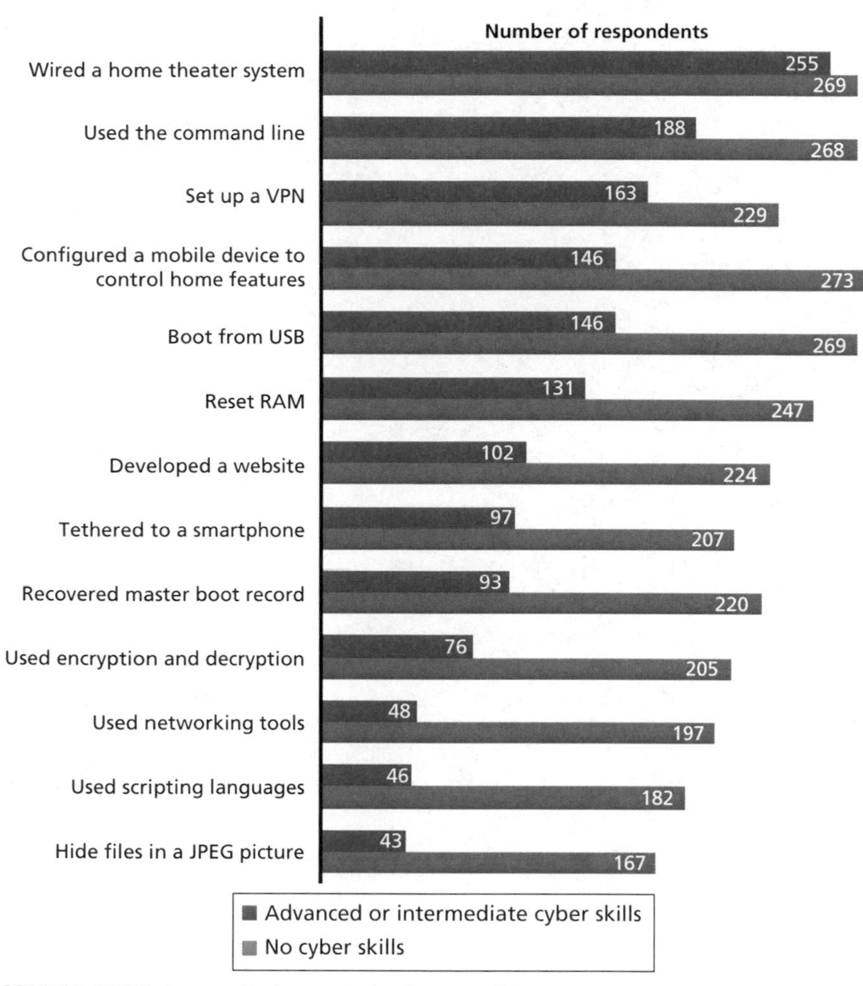

SOURCE: RAND Arroyo Center analysis of survey data.
RAND *RR1490-7.15*

CHAPTER EIGHT

Framework for Examining Current and New Uses of the Reserve Component

Although both AC and RC organizations and soldiers are able to fulfill a full range of cyber roles and missions, this does not mean that each is equally well disposed or composed to conduct any particular cyber task or set of tasks. A given task's characteristics—such as the location of performance—can have a substantial effect on whether it *should* be performed by either AC or RC organizations and/or civilian personnel. The characteristics of some tasks could make those tasks more appropriate for (or "lean toward") either the AC or the RC.[1] For instance, both AC and RC organizations can "provide cyber support." The choice of which organization *should* provide cyber support is determined in part by needed response times, network ownership, and personnel requirements. One could argue that RC units should train in order to deploy and perform AC missions just like the rest of the Army. From this perspective, there is not as much of a need to distinguish between tasks. However, this chapter seeks to identify those tasks that could be the responsibility of the RC.[2]

[1] Note the significant distinction between AC and RC forces in terms of their "fit" for conducting cyber tasks as regards authorities (e.g., Titles 10 and 32) and their availability (i.e., whether forces are on duty or would have to be activated). In the latter case especially, due to the nature of cyber defense and other cyber activities, 24-hour availability and rapid response are requisite capabilities. For the portions of the RC that are not "active," these capabilities cannot be assumed.

[2] R. Wayne Dudding, former commander of the Army Information Operations Group, personal communication with the author, January 31, 2016.

This chapter analyzes select tasks that cyber forces are expected to perform. This is motivated by the need to clarify task performance responsibility in order to inform cyber force requirements decisions across DOTMLPF-P (doctrine, organization, training, materiel, leadership and education, personnel, facilities, and policy) domains. Accomplishing this analysis requires

1. establishing a representative set of cyber tasks, organizing them into groups based on shared characteristics or functions, and characterizing each task as either institutional or operational in nature
2. developing an analytic framework based on a set of fundamental task characteristics to assess each representative task
3. applying the analytic framework and build decision/flow charts for selected tasks to determine whether the task should be conducted by (or lean toward) the AC or RC.

Identifying and Organizing a Representative Set of Army Cyber Tasks

Currently, there is no one universally accepted list of Army cyber tasks for us to reference.[3] In part, this is because the cyberspace domain is relatively immature, but it is also because many of the tasks that would have been previously defined as either "intelligence" or "signal" are now falling within or are being reassigned to the cyber domain. As DoD continues to define the cyberspace domain, we can expect to see a parallel effort to create new cyber tasks and modify existing ones.

The tasks that RC units might perform should not necessarily be viewed as part-time work but perhaps as missions for units to perform when mobilized. In lieu of a reference list of cyber tasks, we instead developed a representative list (Table 8.1) gleaned from a literature

[3] The KSAs of the Cyber Mission Force have a guiding document as discussed in Chapter Five.

review that examined relevant doctrine, briefings, and reports in which cyber tasks are enumerated.[4]

Our literature review produced dozens of closely related cyber tasks that could be grouped together by function (e.g., defend, assess, attack, support, collect, or other)[5]—functions that are closely related to how cyber operations are defined in doctrine.[6]

After developing this list and grouping the representative tasks, we then categorized each task based on whether an institutional or operational organization would likely—by the nature of the task—be responsible for the task's conduct.[7] Tasks that involve defending or attacking cyber assets should belong to the operational units, whereas tasks that involve assessments or providing personnel and support should, generally, belong to institutional units.

---

[4] This review included but was not limited to an examination of the Universal Joint Task List (Joint Electronic Library, *Universal Joint Task List*, April 2015), the Army Universal Task List (Field Manual 7-15, *Army Universal Task List*, Washinghton, D.C.: Headquarters Department of the Army, February 2009), and Section 933 of the Fiscal Year 2014 National Defense Authorization Act (Public Law 113-66), all of which enumerate, in varying detail, lists of cyber and cyber-related tasks.

[5] "Other" consists of tasks that are not purely doctrinal cyber operations but are closely related.

[6] JP 3-12R (2013) defines cyber operations as the conduct of DCO, OCO, and Department of Defense Information Network (DODIN) operations:

> Offensive Cyber Operations (OCO) are cyber operations (CO) intended to project power by the application of force in and through cyberspace. Defensive Cyber Operations (DCO) are CO intended to defend DOD or other friendly cyberspace. DODIN operations are actions taken to design, build, configure, secure, operate, maintain, and sustain DOD communications systems and networks in a way that creates and preserves data availability, integrity, [and] confidentiality, as well as user/entity authentication and non-repudiation.

[7] Institutional organizations function primarily to support operational organizations by providing the infrastructure and support necessary to ensure the readiness of all Army forces. Key institutional tasks include recruiting, training, education, developing leaders and doctrine, providing logistics and facilities support, and ensuring resource optimization. Operational organizations function primarily to perform missions that support the accomplishment of the Army's worldwide strategic, operational, and tactical objectives. For discussion, see U.S. Army, "Organization," web page, undated-a.

This is significant because, although both AC and RC organizations *can* conduct both types of tasks, the RC is relatively better composed for and disposed to conduct institutional organizational tasks, whereas the AC is relatively better composed for and disposed to conducting operational organizational tasks. This is largely because the conduct of many operational organizational tasks requires consistent and/or persistent engagement among cyber organizations, agencies, or entities, and quick or even immediate response times. Both of these requirements are best fulfilled by AC organizations, or at least those that are on continuous active status.[8]

Developing an Analytic Framework

To further refine roles and missions for the Army's RC forces, we developed a set of seven fundamental task characteristics that define what component of the force *should* be conducting a particular task:

1. **Access:** Do cyber forces have the authority to conduct the task (i.e., is the authority for the task's conduct contained in Title 10 or Title 32)?
2. **Risk:** What is the risk associated with the task not being conducted (e.g., probability and consequence)?
3. **Terrain:** What is the cyberspace locus of the task being conducted (e.g., the continental United States [CONUS] or outside CONUS [OCONUS],[9] federal or state)?[10]
4. **Periodicity:** Is the task's occurrence predictable (e.g., periodic or aperiodic)?[11]

---

[8] RC units support the Army Computer Response Team in SWA, which is outside CONUS.

[9] The only OCONUS Army Computer Emergency Response Team mission is being performed by RC units.

[10] Arguably, this is often difficult to distinguish in cyberspace.

[11] Predictability is ideal for reserve unit deployments because a unit will have years to prepare for deployment.

5. **Immediacy:** How quickly must the task be conducted (e.g., near real time or long lead time)?[12]
6. **Frequency:** How often must the task be conducted (e.g., often or rarely)?[13]
7. **Resources:** How many people and with what qualifications, experience, and/or knowledge are required (e.g., high number or low, specialization or no specialization)?

Given any particular task's characteristics and the location/ownership where the task is to be conducted, we can discern whether the task leans toward the AC or RC in terms of who *should* be best disposed to or composed for its conduct. For example, not all RC operations are authorized under Title 32; they can be conducted under Title 10.[14] If the task requires Title 10 authorities, or if the location of its conduct is on a federally owned system or network, it leans toward the AC, but if it requires Title 32 authorities and is conducted on a system or network at or below the state level, it leans toward the RC. Emphasis is placed on the word *lean*, since, again, not all RC operations are authorized under Title 32.

The notional example in Figure 8.1 shows a decision chart that considers a task's risk, periodicity, and frequency. Ultimately, this kind of analysis can inform decisions about which component, AC or RC, *should* be performing which tasks. Continuing this kind of analysis for the range of cyber tasks that Army forces are expected to perform

---

[12] Often, cyber is an ongoing mission although crisis events may require the need to selectively recall individuals.

[13] According to Dudding (2016), an Army Force Generation (ARFORGEN) cycle of 1:5 would work well for needs that are infrequent or rare.

[14] As noted by LTG Jeffrey Talley, 32nd Chief of Army Reserve, in his 2016 Posture Statement,

> [t]he National Defense Authorization Act of 2012 expanded the ability of the Army Reserve to assist in domestic emergencies. Section 12304a of title 10 U.S. Code allows the Army Reserve to provide life-saving, life-sustaining capabilities for Defense Support of Civil Authorities (DSCA) for up to 120 days in response to a Governor's request. (LTG Jeffrey W. Talley, *The 2016 Posture of the United States Army Reserve: A Global Operational Reserve Force*, submitted to the U.S. House of Representatives Appropriations Committee, March 22, 2016)

Figure 8.1
Example Decision Chart

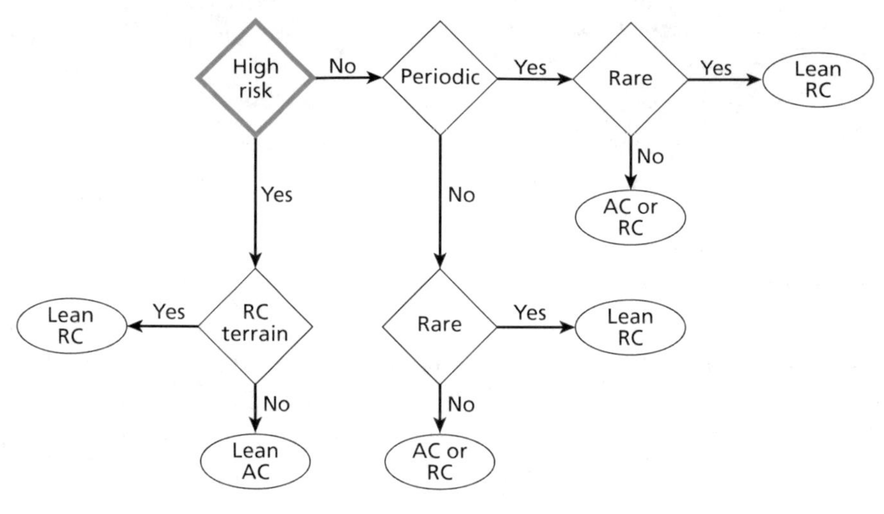

RAND *RR1490-8.1*

could also help inform decisions about force requirement across the DOTMLPF-P domains.

As the figure shows, there are high-risk networks that demand 24/7 attention that fall into the responsibility of the RC, notably ICS SCADA, Corps of Engineers networks, and the Army's MEDCOM net, and Guardnet.

Policy and Authorities Have a Significant Effect on the Conduct of Tasks and, Hence, on Cyber Roles

The seven characteristics enumerated earlier can help define which way the conduct of particular cyber tasks *should* lean (toward either AC or RC cyber organizations). Although we did not rank these characteristics in terms of their importance or value, we did note that certain characteristics—particularly authorities and where the network is located—can disproportionately affect whether an AC or RC organization is, under normal circumstances, permitted to conduct or prohibited from conducting a given task. With respect to access, existing policy and U.S. law heavily influence the possible roles that the AC or RC can play in the conduct of cyber tasks. For this reason, in the minds of some in DoD and elsewhere,

> it remains unclear what role the U.S. military should play in defending U.S. companies and critical infrastructure against cyber attackers . . . many in the military are reluctant to assume responsibility for defending nonmilitary cyberspace [due to] legal restrictions against military involvement in domestic intelligence gathering and law enforcement.[15]

It should be noted that the specific role of the National Guard in the conduct of cyber operations is being examined from policy and authority perspectives by the Office of the Secretary of Defense–Policy (OSDP) and others.

Understanding authorities issues will be critical to the employment of the National Guard in cyber operations. Currently, the National Guard frequently operates as a "state-directed" force instead of acting overtly under Title 32 (see Table 8.1). This is primarily because OSDP is hesitant to allow invocation of Title 32 authorities for cyber operations. The concept of life and property risk is important to the invocation of the title and could extend logically to include electric power, water, food, railway, gas pipelines, and so forth. One consideration to note is that in cases where a cyber attack against the United States is successful (or even partially successful), a life and property emergency is also concurrently likely. Thus, the probability of a strictly cyber incarnation of this same emergency seems low.

Authorities issues also affect interstate assistance. For example, under Title 32, a National Guard soldier in South Carolina would not be legally allowed to assist in the conduct of a cyber mission in the neighboring state of Georgia. And while a joint task force could reach across state boundaries, Title 10 authorities are required for these organizations to work more closely together in the event of a cyber-

---

[15] William Matthews, "Growth Mission," *National Guard Magazine*, Vol. 66, No. 6, June 2012.

Table 8.1
How the National Guard Can Be Activated

What	Who Orders This	Who Pays for This	Duration
State active duty	The state governor upon declaration of an emergency	The state	
Title 32 full-time National Guard duty		The federal government	Individuals can be involuntarily activated for up to 24 months (during a 6-year period)
Inactive duty for training (IDT)			Proposals for extending from 2 to 7 weeks per year. Some units already spend 6 weeks mobilized
Title 10 federal duty	The President or Secretary of Defense with the approval and consent of the state governor	The federal government	

SOURCE: Wikipedia, "National Guard of the United States," August 2015; U.S. Army, "Military Leave for Inactive Duty Training," June 6, 2006.

related contingency.[16] In each of these examples, proper authorization can substantially limit who is allowed to conduct what tasks and where the task may be conducted. As LTG Cardon noted in his testimony, "authorities are a complex problem . . . every state is different."[17]

Using the National Guard to conduct cyber operations is further complicated by who or which agency is permitted or required to activate units, the duration of the activation, and who is responsible for paying for a unit's activation. Different means of activating the National Guard, how, and for what time period are shown in Table 8.1.

It should be noted that states can create and have created relationships through and are protected by nondisclosure agreements with

---

[16] David Halla, U.S. Cyber Command/J7, personal communication with the authors, April 25, 2014.

[17] Cardon, 2015.

private contractors, but if they are under DHS authority, additional complexities arise. State governors often attempt to fix the problem before requesting assistance. This is because states want to retain the capacity to respond to events first while avoiding the possibility of the federal government deprioritizing them in the event of an emergency.[18]

Past and Present Role of the USAR

As articulated by LTG Jeffrey Talley, quoting the 2016 Posture Statement:

> The Army Reserve committed more than 800 soldiers directly, and 3,500 soldiers indirectly to support cyber operations. . . . These 3,500 soldiers come from signal units that provide defensive cyber operations support DODIN. These 3,500 positions supporting signal cyber operations encompass soldiers assigned to perform a cybersecurity mission set. The 1,545 by FY2016 represent those assigned to cyber units performing cyber as their primary mission. The rest encompass the signal soldiers assigned down to the unit level who perform their cybersecurity mission in support of the overall DOD information network.[19]

Other Potential Roles of the RC

Broad potential roles for RC cyber forces are enumerated in Table 8.2. We built this table using data from U.S. Army Cyber Command briefs and our own inclusions. Tasks that are institutional in nature, under normal operating conditions, are good candidates for RC cyber units. These include conducting assessments, providing expert advice, and supporting exercises. Operational tasks, especially those involving the defense of RC networks, are included.

---

[18] Halla, 2014. Fifty-four states and territories have Emergency Management Assistance Compact (EMAC) agreements regarding the sharing of personnel and resources.

[19] Talley, 2016.

Table 8.2
Possible Roles and Mission for the U.S. Army's RC Cyber Forces

Military Mission	Cyberspace Action	Potential Requirements for RC
DCO	Cyber defense	**Defend:** Army ICS SCADA, Corps of Engineers Network, Army Labs and FFRDC networks, CADET command networks, medical command networks, Guardnet
		Regional support command networks, RC training environments, Army Reserve Networks, CONUS ARNG division and BCT networks
	Cybersecurity	**Assess:** critical infrastructure, e.g., Critical Infrastructure Risk Management, Critical Infrastructure Protection; regional support command networks; ARNG Division and BCT networks and vulnerability of C4ISR systems; risk of supply chain; risk on state networks; emergency management networks; command cyber readiness inspections **Advise:** Technical support/advice to state, regional, local governments; integrate private industry knowledge
OCO	Cyber attack	Targeting; gaining and maintaining access
General cyber support		**Exercise and training support:** provide combat training center with opposing forces, provide cyber instructors, provide a training pipeline, establish education and training plans; cyber threat emulation **Surge capacity:** Provide cyber planners, provide cyber analytic team, provide vulnerability assessment teams, provide legal/staff judge advocate support, provide personnel for the Cyber Mission Force
Cyberspace information collection		Support state intelligence fusion centers; theater cyber PIR and IR
DODIN		Provide RCERT, CND-SP; secure and operate Guardnet; securely operate Army Reserve Net
Electronic warfare		Conduct electronic warfare

We do not describe any U.S. Cyber Command teams that could be involved with the military missions, although we acknowledge they will certainly play a large role.

Chapter Summary and Conclusions

Task characteristics are important when determining the roles and missions of AC and RC cyber organizations. *What* a cyber organization is expected to do and *when* and *where* it is expected to do it are of substantial import to whether a task should be conducted by AC or RC organizations. While we were unable to conduct specific task analysis—that is, including explicit mission conditions (such as location where the task is going to be conducted, owner of the network of the affected system, and so forth) and task characteristics—during the course of this study, we were able to develop a representative list of tasks and characteristics that could be used as the starting point for additional analyses using the analytic framework presented above. We believe the framework we have developed in this chapter is useful. However, it will be more difficult to fully define all of the tasks that fit into the framework. This is left as future work.

Discussion: Efficient Use of Highly Skilled RC Personnel

Figure 8.1 assigns tasks and personnel based on a set of decisions that do not consider the efficiency of using highly skilled, highly talented RC personnel on tasks that do not have the same level of required complexity and expertise. It would not be efficient to use highly talented RC personnel for unimportant (low-risk, infrequent, rare) tasks, and some would argue that, ideally, individuals or teams should do work that is aligned with their skill sets. We do not attempt to address this in our decision framework, but we acknowledge that there is the need for future discussion.

CHAPTER NINE

Reviewing the Army's Cyber Human Capital Strategy

Cyber operations require talented and knowledgeable personnel. As discussed in the previous chapters, many organizations, including those in private industry, need the same type of personnel, and the Army finds itself in direct competition with them. In spite of the contention, the Army's requirements for personnel to support cyber operations continue to grow. Specifically, the Army must contribute personnel and units (CPTs, cyber mission teams, etc.) to the Cyber Mission Force.[1] It must also develop its own forces dedicated to Army missions, such as securing Army networks. For these reasons, a comprehensive human capital strategy is a necessity and is currently an ongoing development. In this chapter, we develop additional recommendations to contribute to that discussion. We do this by reviewing what the Army has already defined as its approach and comparing this to the strategy of others. We then assess the completeness of the Army's approach with respect to the RC.

[1] As noted by Tice, "The Cyber Mission Force will have both offensive and defensive capabilities, and will be part of a multiservice force structure of 133 cyber mission teams coordinated by U.S. Cyber Command" (Jim Tice, "Staffing Goal for Cyber Branch Totals Nearly 1,300 Officers, Enlisted Soldiers," *Army Times*, June 15, 2015b).

The Army's Current and Future Goals

The Army is planning to develop a cadre of cyber operators using a number of approaches, including a new branch and new incentives. The Army has an initial goal of assembling a team of 355 officers, 205 warrant officers, and 700 enlisted soldiers.[2] The assessment will be done mostly from the AC through "ongoing accession, branch transfer and reclassification options."[3] Longer-term goals include expanding into the RC.[4]

What the Army Has Done to Reach Those Goals

The Army strategy includes recruiting "from within its own ranks" by creating a new cyber branch via a new career management field. Other incentives are being employed to attract and retain personnel toward a cadre that makes cyber a military career.[5]

Established a New Branch

The new 17-series cyber career field (i.e., career management field 17) will help manage professional growth of the Army's "cyber warriors." The branch has already accepted officers that are focused on cyber warfare, and many of these have been assigned to the 780th Military Intel-

[2] Tice, 2015b.

[3] "The Army expects to reach an 80 percent strength level for the enlisted component of Cyber Branch by 2016" (Tice, 2015b).

[4] According to LTG Cardon, U.S. Army Cyber Command will create "a total, multi-component Army cyber force that includes 21 Reserve-component cyber protection teams, trained to the same standards as the active-component cyber force" (quoted in David Vergun, "Cyber Chief: Army Cyber Force Growing 'Exponentially,'" army.mil website, March 5, 2015).

[5] Kevin McCaney, "Army's New Cyber Branch Looking to Recruit Talent," *Defense Systems*, December 11, 2014.

ligence Brigade. The goal of the branch is 1,300 officers and enlisted personnel.[6]

Quoting the Army's Cyber Commander:

> The establishment of a Cyber branch shows how important and critical the cyber mission is to our Army, and allows us to focus innovative recruiting, retention, leader development, and talent management needed to produce world-class cyberspace professionals.[7]

In addition, the 25D military occupational specialty (named the "cyber network defender") is available for certain noncommissioned officers,[8] and the 255S (named "cyber defense technicians") is available for warrant officers. These personnel currently have designated roles in both Cyber Mission Force teams and Army brigades to defend brigade networks.

Created Bonuses and Incentives

The Army's Human Resource Command has established a cyber-specific selective retention bonus.[9] According to the Army, for qualified personnel, e.g., 35Qs with additional skill identifier (ASI) E6, "bonuses will range from $12,300 to $50,400 depending on grade and service commitment. For 35Q EAs, bonuses will range from $7,900 to $32,200."[10]

---

[6] David Ruderman, "Army Offers Selective Retention Bonuses to Retain Enlisted Cyber Warriors," army.mil website, May 29, 2015; Tice, 2015b.

[7] Quoted in Fort Gordon Public Affairs, "Army Cyber Branch Offers Soldiers New Challenges, Opportunities," army.mil website, November 24, 2014.

[8] "There are more than 700 25D positions across the Army, and the MOS is open to experienced soldiers in the grades of staff sergeant to sergeant major" (Michelle Tan, "Army Activates its First Cyber Protection Brigade," *Army Times*, September 9, 2014).

[9] According to the Army, the bonuses will "initially impact Soldiers in their military occupational specialty, or MOS, 35Q, cryptologic network warfare specialists, with an additional skill identifier E6 (interactive on-net operator), and 35Q EAs (exploitation analysts) already working with specific cyber units" (Ruderman, 2015).

[10] Ruderman, 2015.

There is already a reclassification bonus in place for enlisted personnel. Specifically, "$4,000 bonuses are available to sergeants and staff sergeants who reclassify to MOS 25D (cyber network defender)."[11]

Other incentives tied to the specialties in Table 9.1 are as follows:

- The new Information Dominance officer category created better promotion opportunities—e.g., 17A cyber officers compete within the same pool for promotions.
- Warrant officers (e.g., in the 170A role) will have opportunities to earn college degrees (undergraduate and graduate).
- Warrant officers and enlisted soldiers may receive Assignment Incentive Pay (AIP).

Utilized a Voluntary Transfer Incentive Panel

A voluntary transfer effort was announced in 2014 and implemented with a degree of success. A year ago, more than 700 applications were reviewed from AC Army officers (2nd lieutenant through colonel) who wanted to transfer to the 17A cyber warfare officer series and met the

Table 9.1
Select Specialties in the Army's New Branch

Branch Specialty Code	Name	Who	How It Will Be Populated
17C	Cyber operations specialists	Soldiers from private first class to master sergeant	Initial entry and current soldiers transferring
17A	Cyber warfare officer	Officers	Transfers from recruiting cryptologic network warfare specialists into 17C roles
170A	Cyber operations technician	Warrant officers	Transferring personnel from military intelligence and signal warrant officers into 170A roles

SOURCE: Ruderman, 2015.

[11] Jim Tice, "Reclassification Cash for Sergeants, Staff Sergeants," *Army Times*, May 6, 2015a.

clearance requirements.[12] Developments continue, including a 2016 panel convened to further "identify soldiers with pre-existing skills needed to perform cyber missions to provide training equivalency for those skills."[13]

Established a Cyber School at Fort Gordon

According to the Army,

> the newly created Cyber School at Fort Gordon, Georgia, will train its first class of lieutenants this summer, followed by captain, warrant officer and noncommissioned officer courses in 2016, and the first Advanced Individual Training course for privates in early 2017.[14]

The Army Cyber Center of Excellence at Fort Gordon is the focal point of the school and most Army cyber training.[15] The Army's Cyber School has indicated that it is working with the private sector through industrial partnerships and the use of cybersecurity certifications to train and vet candidate personnel.[16] The Army is implementing

---

[12] The requirement:

> [i] Be able to obtain a top secret security clearance with access to sensitive compartmented information. The clearance must be obtained before attending the 17A qualification course. [ii] Be able to obtain and maintain a counterintelligence polygraph exam, and have access to National Security Agency facilities. NSA [National Security Agency] access will be required for many assignments with the Army's cyber mission force. (Jim Tice, "Officers Can Apply to Go Cyber in Voluntary Transfer Program," *Army Times*, October 6, 2014)

[13] U.S. Army Cyber Command, "Summit Brings Senior Cyber Leaders Together to Share Total Army Opportunities, Solutions," army.mil website, January 5, 2016.

[14] Tice, 2015b.

[15] The Cyber School engages in four key activities: building the school, building the branch, training the force, and establishing the culture.

[16] These partnerships are organized into three divisions: the network management division, the information dissemination management division, and the cyber-electro-magnetic-activities division.

a phased approach to building out cyber capabilities throughout the branch.[17]

Established the Army Cyber Institute

The Army Cyber Institute located at West Point has been producing cadets who are focusing on studies of cyberspace operations. This includes cadets who will become cyber branch officers. The institute is also looking to expand to Reserve Officers' Training Corps (ROTC) programs. In addition, they have done some work in defining what a career development model might look like, created the cyber branch insignia, and are supporting the Cyber Center of Excellence and Army Cyber Command.

What Others Have Done in Terms of Cyber Human Capital Strategies

U.K. Reserve Model for Cyber

The United Kingdom employs a cyber unit within its reserves.[18] The group is called the Land Information Assurance Group (LIAG) and has operated and trained all over the world.[19] The LIAG consists of highly trained cyber specialists who have already honed their IT skills in industry.[20] Recruits are sought to conduct web application testing, database security testing, vulnerability assessments, computer forensics,

---

[17] In phase 1, workforce roles are being defined with the strategic mission of protecting computer systems and networks associated with DoD and critical infrastructure. In phase 2, electronic warfare workforce roles will be defined, with the tactical goal of enabling commanders to plan and execute the full range of cyber operations. In phase 3, support for a cyber effects electronic codebook will be develop so that cyber and electronic warfare forces can support "Big Army" requirements.

[18] The Army reserve in the United Kingdom was formerly called the National Territorial Army.

[19] For example, Afghanistan, Ascension Island, Brunei, Canada, Cyprus, the Falkland Islands, Germany, Gibraltar, Iraq, Kenya, Kosovo, and the United States.

[20] Central Reserve Headquarters (Royal Signals), homepage, 2015.

firewall testing, and intrusion detection and serve as security architects and network traffic analysts.[21]

The strategy for maintaining cyber proficiency is to rely on the industry affiliation of the reservists to provide all the training needed to stay abreast of new developments. For this reason, the reservists are required to have at least five years of "practical IT security experience" and the corresponding professional certifications. To better enable this, reservists can join at age 50. The qualification process is strict and includes two different board reviews. Most LIAG members make high salaries when they are not in uniform, working as highly skilled cyber specialists. Reservists are required to be able to serve for at least 19 days a year but often serve for longer periods that could last four continuous months of deployment.[22] In the LIAG, reservists become officers regardless of their educational background. Reservists can come from other military units or without any military background.

Use of Cyber Aptitude Tests

The U.S. Air Force employs a cyber aptitude test, as does the Israeli Defense Force.

Other U.S. Service Branches

All the services' branches have their own cyber commands (e.g., the Navy's Information Dominance Command), personnel strategies, and plans to use reservists for cyber missions.[23] Previous impediments to

---

[21] From Central Reserve Headquarters (Royal Signals) homepage (2015):

> Non-Commissioned Recruiting–LIAG are currently recruiting technical specialists with practical experience in the following areas: TEMPEST Testing; Technical Surveillance Countermeasures; Defensive Monitoring; Defensive Internet Monitoring. Requirements: Aged 18 to 50—waivers may be available on the upper age limit; Professionally qualified and experienced in a relevant area; Hold UK or Commonwealth Citizenship; Serve at least 19 days per year; Are willing to undergo Her Majesty's Government Security Clearance to Developed vetting (DV) level; Have lived in the UK for the last 5 years.

[22] The 19 days represent the minimal camp and weekend requirements.

[23] Reserve Forces Policy Board, *Report of the Reserve Forces Policy Board on Department of Defense Cyber Approach: Use of the National Guard and Reserve in the Cyber Mission Force*,

promotion opportunities for cyber specialists are being addressed.[24] Also noteworthy are the service-associated organizations' novel recruiting efforts, which extend to high school and middle school students.[25] It should be noted that the Army is beginning efforts to improve the recruiting of high school students toward STEM subjects.[26]

Federal Agencies

To summarize an article on Nextgov.com attributed to the NSA, federal agencies are facing cyber talent management challenges that are similar to those faced by the Army. In general, the federal government is experiencing attrition among federal employees. Less than 25 percent of the federal cyber civilian workforce is under age 30, and over 50 percent of the federal cyber civilian workforce is over age 50; these statistics suggest that the federal government will soon face a significant shortage of cyber employees due to attrition based on retirement. In addition, as with the U.S. Army, the federal government has difficulty competing with the private sector for cyber talent, due to low federal salaries.[27]

RFPB Report FY14-03, August 18, 2014. The U.S. Air Force created the role of cyberspace operations officer (COO) for officers in the Air Force Reserve. COOs are responsible for the planning, supervision, and security of the vast array of Air Force Reserve computer, communications, operations, and tracking systems. To be eligible for the role, applicants must have a bachelor of science in a designated science, technology, engineering, and mathematics (STEM) field and be serving as an officer in the Air Force.

[24] In the Navy, excessive shore duty can be a detriment. The Navy is in the process of implementing its Meritorious Advancement Program with changes that might better reward members of the force in cyber roles that often involve shore duty. In the Army, cyber officers can compete against each other in a separate officer category called the Information Dominance Officer.

[25] For example, the Air Force Association is one of several that have a CyberPatriot program, a national youth cyber program that seeks to inspire high school and middle school students to pursue careers in STEM fields. In addition, the Air Force Academy and the other service academies have a Cyber Team that regularly competes in cybersecurity/hacking contests. Arguably, these activities are helpful to long-term recruiting efforts for this country's cyber force.

[26] U.S. Army, "U.S. Army STEM Experience," web page, undated-b.

[27] Jack Moore, "In Fierce Battle for Cyber Talent, Even NSA Struggles to Keep Elites on Staff," nextgov.com, April 14, 2015.

The federal government is engaged in some innovative talent management activities. In particular, the federal government has instituted the CyberCorps: Scholarship for Service (SFS) program. SFS is a National Science Foundation–funded scholarship program for students who agree to work in information assurance government roles upon graduation.[28]

The NSA is one federal agency that is faring better than most agencies with managing cyber talent. Although the NSA also faces stiff challenges for maintaining sufficient cybersecurity talent, less than 1 percent of its positions are vacant for significant periods of time, and the NSA hires the largest share of SFS graduates (approximately 33 percent). These accomplishments have been achieved even though NSA salaries are considered low by private-sector standards.

The NSA has several notable cyber talent management efforts. It has created 44 Centers of Academic Excellence in Cyber Operations at partner institutions, including Northeastern University, the University of Tulsa, and the Naval Postgraduate School. The institutions teach cybersecurity skills and also serve as recruitment centers. The NSA also funds many student scholarships at a broader set of institutions. The NSA has developed an iPhone App—CryptoChallenge—that tests the user's pattern recognition skills through a series of cryptographs; the goal of the game is to decipher encrypted quotes, factoids, and historical events. CryptoChallenge serves as a public relations tool that aids with recruitment efforts. In addition, the NSA sends recruiters to

---

[28] The SFS program is an example of how academic institutions can help provide trained personnel by means of the formal education process. This program offers scholarships focusing on cybersecurity or information assurance in exchange for a commitment to work in public agencies. Such programs are small in terms of numbers of personnel developed, so this program by itself is not a solution. However, it can be considered successful: These "Center of Excellence" programs have been able to maintain high continuity after commitment; one study found that after the students graduated and honored their commitment of two to three years at the government agencies, they continued to work in the government instead of leaving for the private sector. Interesting and stimulating work was probably why students chose to remain in organizations such as the NSA and Computer Emergency Response Team (CERT) rather than leave their positions at government agencies in pursuit of potentially higher-paying jobs at private organizations. This suggests that such programs can generate dedicated people who can commit to government and military service.

cybersecurity conferences, such as DefCon, that are considered highly credible among the hacker community.

Many of these techniques contribute to a mystique associated with the NSA. For example, producing a respected iPhone app indicates a level of technical sophistication that can draw candidate employees. Several notable NSA alums have founded startups, such as Sqrrl and IronNet Cybersecurity, which has added to the NSA's mystique.[29]

We reviewed the cyber talent management approaches of the NSA with respect to our three talent management focus areas: recruiting, training, and assignment. As already discussed, the NSA is engaged in several innovative recruiting approaches that start early and go to novel sources of talent. In addition, the NSA engages in extensive training, both internally and through external partnerships. We are not aware of especially unique or novel assignment or placement approaches of the NSA.[30]

Language, Medical, and Legal Corps

The next step of the analysis consisted of reviewing three organizations with specialized roles that the U.S. military administers: Defense Language Force Management, the Medical and Dental Corps in the Military Health System, and the Judge Advocate General's (JAG) Corps. These three organizations were selected because they face talent management challenges that are similar to the four challenges associated with cyber talent management. All four of these specialized roles contend with a specialized skill set, deal with difficulty in recruiting civilian expertise, and face strong competition from the private sector. Cyber talent management within federal agencies does differ from the other three in that it deals with a rapidly changing skill set.

---

[29] Kashmir Hill, "The NSA Gives Birth to Start-Ups," *Forbes*, September 10, 2014; Jack Moore, "The NSA's Fight to Keep Its Best Hackers," *DefenseOne*, April 15, 2015.

[30] "NSA does a good job recruiting, but has a difficult retention problem. It's not clear if it's poor internal management or poaching by the private sector. It's worth saying that the NSA isn't the best model of retention or that new retention models might be needed," (Cynthia Dion-Schwarz, manager of Cyber and Data Sciences Programs at the RAND Corporation, personal communication to the authors, December 10, 2015).

We reviewed the cyber talent management approaches of Defense Language Force Management, the Medical and Dental Corps, and the JAG Corps with respect to our three talent management focus areas: recruiting, training, and assignment.

Defense Language Force Management

From a recruiting perspective, Military Accessions Vital to the National Interest (MAVNI) plays a role in accessing foreign national talent that the private sector cannot leverage. This suggests a system that the RC could use to distinguish itself from the private sector if MAVNI was extended to include cybersecurity.

From the perspective of training, the U.S. military engages in both extensive internal and external training. Viewing these training activities with a cyber lens, we see that the role of the Defense Language Institute Foreign Language Center plays a role similar to that of the Army Cyber Center of Excellence, and the role of university partnership training is similar that of to private-sector certifications.

From an assignment perspective, the Defense Language Force Management's systematized assessment process would be useful in cyber talent management, where no such process currently exists.

Medical and Dental Corps

The Medical and Dental Corps engage in several recruiting and retention activities. All medical and dental professionals are commissioned officers, which can influence recruitment (though it is not clear that this would carry over to cyber recruiting, given that titles are not a primary driver for cyber roles in the private sector). The Joint Medical Executive Skills Program facilitates assignment for health care system leadership. As with Defense Language Force Management, MAVNI applies to health care, serving as another example of its potential application to cybersecurity recruiting.

Many of the remaining recruitment efforts address financial compensation. Financial incentives may help in recruiting for Army RC cyber roles, but it is not clear that this tactic would be sufficient without additional techniques.[31]

The Medical and Dental Corps are able to leverage external training by recruiting medical and dental school graduates. In the case of cybersecurity, external training could be based on cybersecurity certifications. In addition, university partnerships (similar to the NSA's Centers of Excellence in Cyber Operations) could provide additional opportunities for taking advantage of external training. The military's Uniformed Services University of the Health Services medical school provides internal training and plays a role that is similar to the Army's Cyber Center of Excellence. External training is beneficial because of the fairly low cost (in comparison with internal training efforts). Overall, there are useful lessons from this community.[32]

JAG Corps

As with health care professionals, all JAG professionals are commissioned officers. JAG recruitment efforts also include several financial incentives (though not as many as for health care professionals).

The whole-person recruiting approach of JAG may offer an innovative way to distinguish cybersecurity recruiting in the Army RC from that of the private sector. Given the advent of geek culture, it may make sense to emphasize the Army RC as a way to apply cybersecurity skills while also staying fit.[33] From a training perspective, the Judge Advocate General's Legal Center and School plays a similar role to that of the

---

[31] The private sector presents formidable competition for cyber talent, due largely to the competitive salary and equity compensation that the private sector can offer. Nevertheless, several organizations (e.g., the NSA for cyber talent and JAG for legal professionals) have been successful in recruiting and retaining cyber talent through strategies that do not depend on strong financial compensation.

[32] Medical recruiting focuses on specific skill sets required by the military community. Recruiting is focused in that career field and officers are shepherded through the program for maximum success in integration. It could serve as the jump point for recruiting for the cyber community. DoD could create a cyber recruiting vehicle much like the one they have in place for medical/dental. It would focus all the talent through one clearinghouse and help localize the problem set (Borras, 2015).

[33] Haya El Nasser, "Geek Chic: 'Brogrammer?' Now, That's Hot," *USA Today*, April 12, 2012.

Army Cyber Center of Excellence. In terms of assignment, the way in which junior JAG professionals are given significant legal responsibility may suggest an approach that could be adopted in the cyber domain. Giving new entrants the opportunity to work on cybersecurity teams tackling problems of national significance, without a long climb up the corporate ladder, may be a selling point.

Table 9.2 summarizes the distinctive approaches that the NSA, Defense Language Force Management, the Medical and Dental Corps, and others take to recruiting, training, and assignment.

Other Ideas: Enhanced Use of Civilians

Idea: A Civilian Cyber Corps

Dennis Dias presents an intriguing option in his 2008 monograph entitled *Partnering with Private Networks: The DoD Needs a Reserve Cyber Corps*. He calls for the creation of a "cyber 'corps' of skilled civilian professionals" in order to reach "non-uniformed civilian professionals" to assist DoD.[34] His proposed corps includes both offensive- and defensive-related operations. One of the benefits of using private-sector personnel, he argues, would be better tracking of the fast-paced technological changes that come out of the private sector. As further justification for such an approach, he cites as a precedent the 2006 *National Security Strategy*, which directly calls for "[d]eveloping a civilian reserve corps, analogous to the military reserves" for disaster relief.[35] Dias does not discuss challenges with respect to the legal status of the operators in the report.

This approach will be most effective (and attractive) if the participants have a defined and agreed-upon role.[36]

---

[34] Dennis P. Dias, *Partnering with Private Networks: The DoD Needs a Reserve Cyber Corps*, Carlisle, Pa.: U.S. Army War College, March 15, 2008.

[35] George W. Bush, *The National Security Strategy of the United States of America*, Washington, D.C.: The White House, March 2006, p. 45.

[36] Dudding, 2016.

Table 9.2
Synthesis of Others' Approaches to Recruiting, Training, and Assignment

Specialized Role	Talent Management Focus Area		
	Recruiting	**Training**	**Assignment**
Language	• MAVNI (up to age 35)	• Defense Language Institute Foreign Language Center University partnerships	• Defense Language Aptitude Battery • Interagency Language Roundtable
Health care	• Commissioned officer status • Financial incentives • MAVNI (up to age 42) • Recruiting from medical/dental school	• Training via civilian medical schools • Training via Uniformed Services University of the Health Services medical school	• Joint Medical Executive Skills Program
JAG	• Commissioned officer status • Financial incentives • Whole-person recruiting • Recruiting from law school (up to age 42)	• Training via civilian medical schools • Training via the JAG Legal Center and School	• Significant responsibility for junior roles
NSA	• Student recruiting (scholarships) • Cyber conference recruiting (e.g., DefCon) • Recruiting via iPhone apps • Mystique	• Centers of Academic Excellence • Extensive internal training	
UK LIAG	• Highly trained cyber specialists from IT industry • No military background required • Officer status • Can join up to age 50	• Reservists maintain cyber proficiency via their industry affiliations	• Variety of roles • Between 19 days and 4 months per year
Others (e.g., Air Force, Israeli Defense Force)	• Cyber aptitude test used to help gauge the ability of personnel to learn cyber and to assist in placement in the cyberspace workforce		

Navy Studies of the Use of Civilians

There are relevant Navy studies regarding the mix of civilians needed and training approaches. These are not available to the general public, but they were summarized in a 2009 briefing by director of the Navy's cyber/IT workforce, Chris Kelsall.[37] Key points from this summary are provided in Tables 9.3–9.5. We do not attempt to validate this analysis and instead present it to the reader as background.

Contract Personnel

Contract personnel play a large role in filling personnel needs throughout the U.S. government, and their use for "cyber jobs" is no exception. One study found that private contractors accounted for 83 percent of the staff in the Office of the Chief Information Officer of the U.S. Department of Homeland Security.[38] However, there has been an expressed desire to reduce reliance on private contractors throughout DoD. According to former Secretary of Defense Robert Gates in his April 2009 budget request,

> Under this budget request, we will reduce the number of support-service contractors from our current 39 percent of the Pentagon workforce to the pre-2001 level of 26 percent, and replace them with full-time government employees. Our goal is to hire as many as 13,000 new civil servants in FY '10 to replace contractors and up to 30,000 new civil servants in place of contractors over the next five years.[39]

Regardless of the extent to which private contractors are relied on, their use does not obviate the need for uniformed personnel and government civilians to "effectively manage [a] blended cybersecurity workforce."[40]

---

[37] Chris Kelsall, "DON IT Workforce," briefing to integrated product team, September 2009.

[38] Partnership for Public Service and Booz Allen Hamilton, *Cyber IN-Security: Strengthening the Federal Cybersecurity Workforce*, Washington, D.C., July 2009.

[39] Jim Garamone, "Gates Lays Out Budget Recommendations," American Forces Press Service, April 6, 2009.

[40] Partnership for Public Service and Booz Allen Hamilton, 2009, p. i. Dion-Schwarz (2015) notes that "The hiring efforts floundered under sequestration."

Table 9.3
Navy Study of Hiring Options for Civilians

Approach	Pros	Cons
Increase awareness of and priority on utilization of recruitment, relocation, and retention flexibilities	• Enhances competitiveness of job offers • Achieves retention objectives • Permits nationwide recruiting	• Incentives are not centrally managed • Resources are at discretion of local hiring manager
Increase awareness of and priority on utilization of direct hire authority for security personnel	• Decreases length of hiring process • Achieves recruitment objectives	• None
Offer bonus pay for employees who achieve desired levels of certification	• Improves retention • Creates parity with other federal agencies already doing this	• Bonus programs are not centrally managed • Resources are at discretion of local hiring manager
Offer employee referral bonuses. Increase the bonuses with referring fully qualified personnel.	• Essentially "free" recruiting • High potential for good organizational fit	• Bonus programs are not centrally managed • Resources are at discretion of local hiring manager
Create a formal, centrally managed Cyber Intern Program	• Success programs exist—acquisition and financial management interns • Strengthens and prioritizes presence at key universities—draws best talent	• Investment required
Create a Cyber Civilian Recruiting Cadre using employees who have graduated from NSA/IC Centers for Academic Excellence	• Sends the best ambassador possible to recruit for Navy—a Navy employee	• No compensation, award, or other recognition program in place

SOURCE: Kelsall, 2009.

Table 9.4
Navy Study of Civilian Education Options

Approach	Pros	Cons
Leverage Federal Cyber Service Scholarship for Service initiative from National Science Foundation	• Scholarships are already paid for through National Science Foundation grants ($50 million) • Recipients are required to serve 2 years in an information assurance role with the government	• No recruiting infrastructure • Lengthy hiring process • No formal, centrally managed internship program to "hook them early"
Directly recruit at NSA-certified Centers for Academic Excellence in Information Assurance	• Existing ROTC footprint at 34 or 94 universities • Synergies with National Science Foundation program—these universities have received 80% of National Science Foundation grant money	• No recruiting infrastructure • Lengthy hiring process • No formal, centrally managed internship program to "hook them early"
Consider partnerships with Intelligence Community Centers of Excellence	• Existing ROTC footprint at 4 of 10 universities • 2 of 10 universities also certified under NSA program	• No recruiting infrastructure • Lengthy hiring process • No formal, centrally managed internship program to "hook them early"
Create a formal program for cyber education through the Naval Postgraduate School and the Air Force Institute of Technology, similar to existing MBA program established in coordination with the Assistant Secretary of the Navy (Financial Management and Comptroller)	• The Naval Postgraduate School and Air Force Institute of Technology are NSA-certified Centers for Academic Excellence • War College programs focused on cyber are relevant to civilians	• Investment required • Application process not in place

SOURCE: Kelsall, 2009.

Table 9.5
Navy Study of Civilian Training Options

Approach	Pros	Cons
Modularize Information Systems Technician of the Future (IToF) team to conduct "data" training for civilians	• Leverages existing training • Utilizes competencies • Eliminates duplication of training establishments	• Investment required • No quota prioritization and management for civilians • Insufficient berthing at Center for Information Dominance
Offer network operations and other continuous learning training for civilians	• Industry modules available (SkillSoft/Netg) • Follows process established during IToF development	• Investment required • No central learning management system for civilians
Leverage joint cyber training opportunities (Joint Cyber Analysis Course, etc.)	• Leverages existing training • Utilizes competencies • Eliminates duplication of training establishments	• Investment required • No quota prioritization and management for civilians • Insufficient berthing at Center for Information Warfare Training

SOURCE: Kelsall, 2009.

Recruitment of Civilians

Studies of whether the military or federal government can recruit sufficient numbers of cyber personnel have mixed conclusions.[41] A report by the Partnership for Public Service and Booz Allen Hamilton claims that the federal government is having trouble finding and recruiting the cybersecurity workforce it needs and cites the following reasons: (1) an inadequate pipeline of new talent, (2) "fragmented governance and uncoordinated leadership [that] hinders the ability to meet federal cybersecurity workforce needs," (3) "complicated processes and rules [that] hamper recruiting and retention efforts," and (4) a "disconnect between front-line hiring managers and [the] government's HR specialists."[42]

[41] Hosek et al., 2004.

[42] Partnership for Public Service and Booz Allen Hamilton, 2009, p. ii.

Chapter Conclusions and Recommendations

The Need to Tap the Civilian Workforce

Effective cybersecurity demands talented and knowledgeable personnel to protect key cyber terrain, e.g., networks. The Army will need to continually adjust strategies for recruiting, training, and qualifying cyber specialists. As part of a broad strategy, the Army should consider multiple ways to tap into the large pool of civilians who have the skills that it needs. Of course, there are some types of cyber skills—specifically, offensive cyber capabilities—that no private-sector experience could parallel.

Any strategy for developing Army cyber capabilities will rely heavily on a civilian workforce, both government and contractor. The USAR and ARNG should provide a significant pool of talent if these components can recruit personnel whose civilian jobs require related expertise.[43]

Human Capital Strategy for Cyberspace Operations

The Army human capital strategy with regard to cyberspace operations needs to be a holistic one that considers each life-cycle human resources function (e.g., structure, acquisition, distribution, development, deployment, sustainment, compensation, and transition).[44] An integrated human capital strategy must address the right mix and distribution for the Army—AC, RC, civilians, contractors, soldiers, officers, and warrant officers. The Army's current human capital strategy

---

[43] The CP3 (Cyber Private Public Partnership) is an Army reserve effort that uses external partnerships with universities and industry in a way designed to bolster the Army's talent management efforts (see U.S. Army Reserve, *USAR Cyber P3*, undated). According to the Army, "the initiative brings together the U.S. Army Reserve, six universities, and nearly a dozen private-sector employers to train "elite cyber warriors" who will serve in the army, elsewhere in the public sector, and in the private sector" (The Intersector Project, "Cyber P3 to Build 'Network of Cyber Warriors,'" February 27, 2015).

[44] Rhett Hernandez, former commander of U.S. Army Cyber Command, personal communication with the author, December 15, 2015.

is new and adopts many of the suggestions from the literature and borrows novel approaches used by other agencies and service branches.[45]

The Army's Cyber Career Field

The Army's creation of new cyber-focused occupational specialties (e.g., career management field 17 and functional areas) is a critical development that demonstrates to commissioned and enlisted personnel that they can have viable careers that are focused on developing the skills and attributes best suited to leading cyber-focused organizations. Most certainly, this will help the Army get better at recruiting and retaining highly skilled cyber personnel.

Borrowing Ideas from Overseas

There are further enhancements that could be made. For one, the U.S. Army, and specifically the RC, should consider adopting aspects of the UK (LIAG) reserve model. For example, the LIAG allows older persons (up to age 50) to join to perform cyber duties.[46] In addition, the LIAG has a selection process that is focused almost singularly on the applicants' technical skills acquired outside the Army. Further, other services and countries are employing cyber aptitude assessments for recruiting, which also makes sense for the Army, whether for the RC or the AC.

Perhaps the Army needs to better consider a way for personnel to move seamlessly between AC and RC service, as is nearly the case for the UK LIAG. This has been discussed outside of the cyber realm, due to enduring conflicts, and is analogous to the medical field.[47]

---

[45] Hernandez, 2015.

[46] According to a 2011 Army brochure, physicians age 47 years or older can join the reserves with a waiver. Those age 46 or lower are also accepted without a waiver (U.S. Army, *Army Reserve Medicine*, RPI 720 FS, May 2011).

[47] Dudding, 2016.

CHAPTER TEN

Main Findings and Recommendations

Advocates of the use of the RC for cyber operations cite a number of reasons. Chief among them is the RC's potential to provide surge capacity for various cyber roles within each service and in U.S. Cyber Command's Cyber Mission Force. In addition, the RC has been suggested as a means of retaining valuable cyber personnel; e.g., recouping DoD's investment in its extensively trained personnel when they leave the AC. For the National Guard in particular, a homeland defense mission is envisioned as an ideal role, especially given the increasing concern over the risk to the nation's critical infrastructure (including the power grid).[1]

Pessimistic assessments of the value of the RC in cyber are influenced by the lengthy training requirements in place today for key roles in the Cyber Mission Force and the possible unavailability of RC personnel to complete this training. However, it remains to be studied whether the DoD training and education regimen needs to include systems other than those most unique to the military (e.g., weapons). It is possible that civilian-acquired training and experience (and proper credit/equivalency) is already sufficient for many of the roles in the Cyber Mission Force.

---

[1] There remain many considerations and unresolved issues for this role. Specifically, to what extent will the guard be permitted to operate on industry networks? Do we know who "owns" the data on these networks? Cyber Guard 2014—a two-week DoD exercise designed to simulate a domestic cyberspace incident—identified this, and other coordination issues, as critical.

Based on both quantitative and qualitative analyses, we find that relevant IT and cyber skills are in abundance in the private sector. As a result, there are tens of thousands of "citizen-soldiers"—that is, soldiers in the Army RC—who have the potential to support the Army's cyber mission needs.

Table 10.1 compares this projected supply with projected demand for cyber-skilled personnel.

Findings

This overarching conclusion was informed by the following findings:

- **Finding 1: More personnel are needed.** According to testimony by LTG Edward Cardon, the Army's current needs are "3,806 military and civilian personnel with core cyber skills."[2] This is more than the Army has now. In the future, the Army's demand will be higher. This projected shortage will be exacerbated by a rapidly growing demand for cybersecurity personnel in the pri-

Table 10.1
Comparison of Projected Supply and Demand for the Cyber Workforce

	Expertise Level	Low Estimate	High Estimate	Source
Projected (potential) supply from the RC	Deep	477	723	Analysis of CEI data shown in Table 4.5
	Mid-level	16,636	18,102	
	Partial	111,626	114,698	
Potential future demand for the Total Army		49,000		Projection based on vendor surveys and industry trends, as shown in Figure 2.6

SOURCE: RAND Arroyo Center analysis of CEI data.

[2] Cardon, 2015.

vate sector. Recruiting from across the Army should be fruitful, but recruiting from outside the Army is unavoidable.

- **Finding 2: Data are available.** The level of cyber expertise that exists in the RC can be estimated with the currently available data sources. This includes the CEI database, the WEX database, and, potentially, novel uses of social media, such as LinkedIn profiles.[3]
- **Finding 3: More detailed data could be useful.** DoD and the Army would benefit from a more detailed inventory of their cyber professionals, relative to what is provided in existing data. The self-reporting effort captured in the CEI database can be improved by adding questions that are more specific to the relevant subareas and specific skill sets of the cyberspace workforce. Specifically, there should be questions that clearly articulate the definition and standard of each cyber skill.
- **Finding 4: Basic cyber skills to support the nation's cyber mission force can be acquired in the private sector.** Most (but not all) of the KSAs for cyber operations—specifically, those identified by the U.S. Cyber Command as requirements for many of the roles that support Cyber Mission Force—can be "civilian-acquired." The private sector has cyber skills and training, and the RC can help utilize personnel with this training. Many of these KSAs can be acquired in the private sector via civilian-based training and experiences. Specifically, they can be acquired in part from popular certificate programs (e.g., CEH, CISSP, Security+) and civilian-sector on-the-job training.
- **Finding 5: Sufficient operations tempo is vital to stay "cyber-sharp."** Many RC personnel are employed in leading-edge technology companies and have critical skills and experience in fielding the latest IT systems, networks, and cybersecurity protocols. Arguably, their nonmilitary employment allows them to more easily maintain currency in their cyber skills, compared with

[3] A 2015 snapshot of current employers of LinkedIn users showed 18,410 for USAR and 30,851 for ARNG. This is a small but a significant fraction of guardsmen and reservists.

some AC soldiers who are not engaged in cyber tasks on a frequent basis.[4]

- **Finding 6: Many cyber-experts in the RC are underutilized.** There are personnel in the RC whose civilian cyber expertise is not being utilized in or applied to their Army careers. This untapped cyber potential is approximately 11,000 people who, at a minimum, have the propensity to learn the cyber skills needed for Army cyber operation.
- **Finding 7: Many RC personnel are interested in working in a cyber field.** There are indications that many in this pool of untapped cyber potential have a desire to use their cyber-related skills in the Army. Many others who do not have cyber skills have a strong interest in acquiring them.
- **Finding 8: The Army will need to continually adjust its strategies for recruiting, training, and qualifying cyber specialists.** Potentially effective options for reserve recruiting include the use of expanded age ranges and generous compensation for sufficiently trained personnel in the private sector.[5]
- **Finding 9**: **The Army should utilize a cyber aptitude assessment tool**, similar to what the Air Force, the National Security Agency, and other countries utilize,[6] to aid recruiting for cyber personnel.

Recommendations

Proceed with the Incorporation of RC Personnel into Plans for the Army's Cyber Force

The opportunity to utilize personnel with training and experience enhanced in industry suggests that the ARNG and USAR are ideal sources of cyber talent. There is sufficient overlap between the KSAs

---

[4] Ultimately, this depends on actual operations tempo and relevance of day-to-day IT skills.

[5] See Table C.3 in Appendix C for specific examples.

[6] Nicole Blake Johnson, "The Air Force Has a Plan for Testing Cyber Aptitude," govloop.com, August 18, 2015; Reed, 2015.

required for the Cyber Mission Force and those used in the civilian IT industry to suggest that there is value in the pool of talent employed there. The Army should leverage this pool to the maximum extent possible. At one time, the Air Force set up reserve units near Redmond, Washington, to take advantage of talent working for IT-focused companies, such as Microsoft, in that state. The Army Reserve Cyber Operations Group has had subordinate elements aligned with technology centers since its inception. The ARNG also took similar measures.[7] More-advanced concepts, such as a "civilian cyber corps," make sense but, to a certain extent, can be achieved using the ARNG and USAR forces today. Furthermore, some roles that are offensive in nature demand uniformed personnel, especially if a presence in a theater of operation is required. A good example of the use of Army reservists (e.g., ARCOG) is the support provided to the RCERT-SWA.

Increase Compliance with and Revise the CEI Questionnaire

We recommend that DoD find ways to increase the compliance with the annual CEI questionnaire, perhaps by issuing more-frequent reminders to RC personnel regarding this mandatory task. The questionnaire has the potential to be a great source of data and yield updated analyses on the cyber skills resident in the RC.

The CEI questionnaire should also be modified to ask for greater detail with respect to cyber-related skills. It should allow for specific questions about the type of IT work, so that DoD can discern whether or not the skills being held fall into the categories in Department of Defense Directive 8140.01, *Cyberspace Workforce Management* (e.g., IT workforce, intelligence workforce).

Develop a New Strategy to Manage the Future Cyber Workforce

Army Human Resource Command (HRC) will need different processes and technologies than are utilized today to manage the cyber

[7] According to Borras (2015), "Western element of ARCOG has recruited from and liaised with Silicon Valley companies. North Central element of ARCOG has been working with CMU/SEI as well."

workforce.[8] The cyber workforce will include new and emerging specialties and function areas, and equivalencies for real-world experiences will need to be continually examined and granted. HRC will need to "institutionalize a systemic continual review process since the KSAs will continue to evolve."[9]

---

[8] Hernandez, 2015.

[9] Hernandez, 2015.

APPENDIX A

Literature Review and Findings from Recent Studies

Human Capital Management for the USAF Cyber Force (2010)[1]

Subject or Policy Question

- Identify and analyze the human capital management issues associated with the creation and management of an Air Force cyber force.

Relevant Findings

- The Air Force's specification of how it will integrate cyber capabilities functionally and organizationally to produce capabilities and effects will ultimately define how it will operate in cyberspace. That refined definition will guide the requirements for cyber human capital in skill and number.
- The Air Force has to meet the challenge to organize, train, and equip its cyber force to successfully prevail in any number of warfare scenarios. Moreover, it must develop its force to effectively

[1] Lynn M. Scott, Raymond E. Conley, Richard Mesic, Edward O'Connell, and Darren D. Medlin, *Human Capital Management for the USAF Cyber Force*, Santa Monica, Calif.: RAND Corporation, DB-579-AF, 2012. Much of the content in this section paraphrases or quotes the report.

confront the increasing use of cyber-based tools and techniques in irregular warfare and counterinsurgencies.
- The Air Force faces an immediate challenge in managing human capital. There is a limited supply of personnel with the requisite skills to compose a cyber force that can deliver the capabilities envisioned by the Air Force.
- The cyber organizations analyzed in this research had two types of positions: those with requirements for skills from traditional specialties (e.g., communications-computer, intelligence, developmental engineering, electronic warfare operations) and those that require an augmentation of traditional specialty skills with skills and knowledge associated with specific capabilities: computer network attack, computer network defense, and computer network exploitation. These positions have "cyber-hybrid" requirements, and they exist for officers, enlisted personnel, and civilians.
- Air Force cyber personnel will need additional technical, legal, organizational, and operational skills, as well.
- Most airmen are developed for these cyber-hybrid jobs through organizationally specific on-the-job training programs. This training results in just-in-time cyber skills for just enough cyber personnel. However, a decentralized, organizationally specific development approach might not be enough to build a sustainable cyber workforce.

Relevant Recommendations

- Establish a more comprehensive concept of operations (CONOPS) that addresses the functional, organizational, and operational integration needed to create highly valued capabilities and describes how the Air Force will operate in and through cyberspace throughout the peace-war-reconstitution spectrum of activities. The revised CONOPS should align Air Force planning with the functional, organizational, and operational complexities inherent in mitigating cyber vulnerabilities and cyber threats and conducting cyber warfare.

- Use the revised CONOPS as a basis for stakeholders to specify Total Force human capital requirements (i.e., for AC and RC forces, Air Force civilians, and contractors). More-comprehensive specifications of cyber operations should add precision to the Air Force's specification of the cyber-based skills needed in the force, its classification structure for cyber skills management, and its identification of the best combination of sources within the Total Force for these skills.
- Establish a lateral officer Air Force Specialty Code (AFSC) as a method to manage cyber skills, particularly for policy, doctrine, planning, and programming jobs that will require people steeped in cyber. Use AFSC suffixes to manage cyber skills within other officer specialties. A lateral-entry AFSC would contribute to quickly building leaders in the cyber domain.
- Continue efforts to retool the enlisted communications-computer specialty into an accession-entry cyber specialty, and use suffixes and special experience identifiers to manage cyber skills in other specialties, such as intelligence. These skill sets within enlisted communications-computer specialties are highly congruent with cyber skill sets in network operations, and this congruency supports the use of an accession-entry specialty.
- Continuously assess the cyber force's sustainability. Cyber capabilities, vulnerabilities, and threats are evolving rapidly. Furthermore, skilled cyber personnel may be attracted to career opportunities in the private sector. To keep pace with these challenges, the Air Force should assess cyber skill requirements routinely to ascertain whether current policies and practices will sustain the force.

Limitations or Caveats

The Air Force's cyberspace concept of operations and organizational structure were still evolving when this research was being conducted. As a consequence, the study was designed to be strategically oriented and comprehensive for broad application depending on the courses of action the Air Force eventually selects.

Additional Notes

Many of the issues highlighted in this study have been more fully addressed since its publication; however, the report's emphasis on a strategic approach to human capital management is still highly relevant. For instance, while the mission, goals, and objectives of the cyberspace organization have been articulated, the details related to the selection, development, utilization, and sustainment still pose human capital challenges within the cyber force.

Employment of Reserve Forces in the Army Cyber Structure (2012)[2]

Subject or Policy Question

- Inform decisionmakers about the cyber workforce and the strengths and benefits of ensuring that the discussion surrounding the development of cyber operations as a key capability for the Army includes the entire force (AC, RC, and civilian).

Relevant Findings

- Although DoD is a significant user of IT, it often lags behind the civilian and corporate sector in the cyber domain, and contracting requirements can inhibit the rapid fielding of IT. RC soldiers increasingly serve as the bridge to close the Army's technology gap by offering current skills and leap-ahead capabilities in the cyber environment.
- Citizen-soldiers employed in leading-edge technology companies have critical skills and experience in fielding the latest IT systems, networks, and cybersecurity protocols. RC members' technical skills are often acquired from the private sector, and their non-

---

[2] Jeff L. Fisher and Brian Wisniewski, *Employment of Reserve Forces in the Army Cyber Structure*, Carlisle Barracks, Pa.: U.S. Army War College, May 2012. Much of the content in this section paraphrases or quotes the report.

military employment allows them to more easily maintain currency in their cyber skills, compared with AC soldiers.
- The RC's local, community-based nature gives its members unique interagency cooperation skills. RC soldiers have relationships with state and local agencies and intimate knowledge of other aspects of the domestic operational environment. Unlike AC members who rotate according to the needs of the service, guardsmen and Army reserve soldiers are frequently recruited into specific units and remain in those positions for a long time.
- Cyber skills are extremely valuable to the military but also fragmented. DoD needs to better catalog currently "hidden" RC soldiers' cyber skill sets and consolidate, organize, and concentrate the employment of cyber personnel.
- Effective use of the guard and reserves will lower overall personnel and operating costs, better ensure the right mix and availability of equipment, provide more efficient and effective use of defense assets, and contribute to the sustainability of both the AC and the RC.
- The AC cannot compete with the private sector in terms of salary or in the amount of training it can provide. As a result, many trained cyber warriors cycle out of the military and move to the private sector.
- Much of the cyber capability in both the AC and the RC is unorganized and fragmented, the training is uneven, and real-world cyber missions are lacking or inadequate.

Relevant Recommendations

- DoD should conduct an inventory of its cyber professionals (regardless of service, branch, or MOS) that includes identification of civilian-acquired skills. The existing CEI self-reporting system must be expanded to better fill this emerging requirement (or a new database must be created). For example,
  - The system needs to be fully integrated with the cyber school houses, and these schools must develop an easy-to-understand and frequently updated military schooling equivalency process

to ensure that military members who have completed civilian-related cyber training, certifications, and experience are granted the equivalent military qualifications, occupational skills, and functional areas.
  - The system must allow for partial credit for civilian-acquired skills and provide a user-friendly interface for RC members.
  - The system must include "experience information," allowing service members to insert more than one current occupation, list previously held occupations, and enter experiential information for each position.
  - Incentives for entering information should be considered, as the most difficult part of collecting this information is getting service members to enter and update data.
- DoD should create robust cyber units in the RC to provide the organization, mission, and billets for former AC personnel who, after building skills at the taxpayer's expense, can be retained in the RC where they are available for recall in the event of an emergency. This initiative will help retain these personnel with critical skill sets in the military inventory.
- To fully utilize the RC, the Army and DoD should develop missions and mission sets that are manned by the RC. These missions should be staffed with RC units providing rotations with minimal full-time manning to maintain continuity. These roles would not only maintain skills to support full-spectrum cyberspace domain operations but also provide support and relief for AC cyber warriors.
- DoD should develop a strategically focused cyber force that fully utilizes the capabilities resident in the RC to provide a more-responsive force for national homeland security. For example, DoD should develop the capability at U.S. Cyber Command to manage and direct the RC cyber forces and act as the Title 10 Training and Readiness Oversight or Training Readiness Authority AC higher headquarters for the RC cyber forces.
- To ensure that the RC is properly resourced and trained for Title 10 missions, U.S. Cyber Command and the RC should ensure that cyber units are missioned and resourced using the same process

for cyber used for any other mobilization. The cyber units should be identified, have a cyber mission and a war trace, and be placed in the Army Force Generation (ARFORGEN) model.
- DoD should integrate and support organizations—such as U.S. Cyber Challenge, Digital Forensics Challenge, and CyberPatriot—into the STARBASE program to support the development of cyber skills.

Comparing Air Force Active and Reserve Forces Conducting Cyber Missions (2013)[3]

Subject or Policy Question

- Evaluate the costs and benefits of alternative mixes of Air Force active and reserve units involved in cyber missions.

Relevant Findings

- Many part-time reservists with civilian jobs in private cyber firms have received advanced training in information technology through their graduate education and on-the-job training. The careers of most regular Air Force officers do not include this training.
- Years of experience is an additional benefit. Air Force RC personnel average seven years of experience, compared with three years for regular Air Force officers.
- Cyber units staffed by only full-time regular Air Force personnel suffer a loss in capability because they do not tap the information technology expertise and advanced experience of part-time reservists.

---

[3] Drew Miller, Daniel B. Levine, and Stanley A. Horowitz, *A New Approach to Force-Mix Analysis: A Case Study Comparing Air Force Active and Reserve Forces Conducting Cyber Missions*, Alexandria, Va.: Institute for Defense Analysis, P-4986, September 2013. Much of the content in this section paraphrases or quotes the report.

- Part-time cyber reservists offer the Air Force another benefit, in that when units need skills not already in abundance, RC personnel can easily acquire them through their contact with private firms.
- Part-time reservists offer an attractive way for the Air Force to obtain personnel with advanced cyber training. Many people with advanced skills in information technology can command high salaries at private firms and are unwilling to accept employment as full-time reservists at government pay rates, but some are willing to serve as part-time reservists.
- Integrating RC and AC cyber units is a way to better utilize the advanced training of part-time reservists. The benefits of integration depend on the mission and can include a higher degree of integration and personnel being able to spend more time on mission tasks instead of education, training, and administration. Whereas personnel in stand-alone traditional RC cyber units spend 25 percent of their time on useful work, personnel in blended units spend up to 60 percent of their time on such work.
- Not all features of integration are positive. For example, associate and blended units have suffered from morale issues.[4]

Relevant Recommendations

- The Air Force might perform the CNA mission with higher value by using an integrated blend of active and reserve personnel.
- The Air Force might lower total personnel cost to the government by encouraging reservists to volunteer for service time beyond the required 39 days.
- The Air Force might improve the performance of both AC and RC personnel by using fully integrated units in which RC personnel work side by side with active personnel in blended rather

---

[4] Borras (2015) states,

> We cannot underscore the importance and value of unit esprit de corps and morale. Successful integration strategies require a good level of inclusion rather than tolerance. TRANSCOM is a great example. They utilize RC during battle assembly and annual training periods in a way that ensures seamless integration of the mission and capability.

than stand-alone operations. This could be particularly important for RC personnel in the CNA area, where operational currency is vitally important and only maintainable by continuing mission work.

Limitations or Caveats

The precision of the assessment is limited by the rating scale used to elicit opinions from subject-matter experts, by some ambiguity in the questions posed to these experts, and by limited access to data. Furthermore, scalability must be further evaluated. Finally, additional manpower may be required in regular Air Force units to maintain the recruiting, training, and support pipeline for active and reserve personnel.

Suitability of Missions for the Air Force Reserve Components (2014)[5]

Subject or Policy Questions

- What considerations affect the suitability of a given mission for assignment to the RC versus the AC of the Air Force? What suitability criteria can be derived from these considerations?
- How can these considerations be used to evaluate the optimal distribution of force structure to the AC and RC for a given mission?

Relevant Findings

- There are three main criteria for evaluating the suitability of missions for assignment to the RC:

---

[5] Al Robbert, James H. Bigelow, John E. Boon, Jr., Lisa M. Harrington, Michael McGee, S. Craig Moore, Daniel M. Norton, and William W. Taylor, *Suitability of Missions for the Air Force Reserve Components,* Santa Monica, Calif.: RAND Corporation, RR-429-AF, 2014. Much of the content in this section paraphrases or quotes the report.

- Surge demand: Force structure is suitably placed in the RC only if there is an anticipated wartime or other episodic surge in demand for forces.
- Duration of activations: Missions with shorter activation periods are more suitable for assignment to the RC.
- Continuation training requirements: Missions with a pronounced continuation training requirement are more suitable for assignment to the RC.

Relevant Recommendations

- Evaluate whether assignment of space and cyber missions to the RC is cost-effective.
- Change the programming and management of Military Personnel Appropriation man-days to include consideration of costs and outputs.
- Seek legislative changes to remove the constraints on the use of technicians, active guardsmen and reservists, and Reserve Personnel Appropriation–funded part-time reservists for duties other than training or administration of reserve forces.
- Adopt more widespread use of cost assessments that consider costs and measured outputs, as well as wider dissemination of cost evaluation results, so that all stakeholders gain a better understanding of how costs for various outputs differ between active and reserve units and how these cost-per-output differences affect the overall costs of various force mixes.
- Review and revise organizational constructs (i.e., classic associations vs. individual mobilization augmenter constructs vs. separate active and reserve squadrons) to improve cost-effectiveness.

Limitations or Caveats

This report centers on the Air Force rather than on the Army.

How the Military Competes for Information Technology Personnel (2004)[6]

Subject or Policy Question

- In light of burgeoning private-sector demand for IT workers, escalating private-sector pay in IT, growing military dependence on IT, and faltering military recruiting, what basis, if any, offered assurance that the supply of IT personnel would be adequate to meet the military's future IT manpower requirements?

Relevant Findings

- The services have been successful in attracting and keeping IT personnel. Compared with non-IT recruits, IT recruits were of higher quality, signed on for somewhat longer terms, had lower attrition, and had similar rates of reenlistment (except in the Army).
- IT training appears to be central to the attractiveness of military IT positions to potential recruits.
- Even if future IT manning requirements change, the military should be able to meet its needs. However, large, abrupt increases in IT manpower requirements will decrease the likelihood of this outcome.

Relevant Recommendations

- Develop a process for conducting broader reviews of personnel force structure and compensation.
- Define and evaluate, by field demonstration, alternative resource configurations, and develop metrics to measure their effectiveness.

---

[6] James Hosek, Michael G. Mattock, C. Christine Fair, Jennifer Kavanagh, Jennifer Sharp, and Mark E. Totten, *Attracting the Best: How the Military Competes for Information Technology Personnel*, Santa Monica, Calif.: RAND Corporation, MG-108-OSD, 2004. Much of the content in this section paraphrases or quotes the report.

- Collect data on the productivity of IT personnel, the underlying factors determining productivity, the value of additional experience in IT, and the optimal experience mix of IT personnel.
- Provide the defense manpower research community with the data it needs to assess whether IT manpower requirements—or IT enlistment and retention targets—are optimal.

Limitations or Caveats

This report was published in 2004, and its findings might be outdated. The findings were focused on the AC.

Additional Notes

The findings and recommendations offer a glimpse into an understudied area of motivations driving IT personnel to work and stay in the military. Most importantly, the allure of IT training appears to be the influential factor.

APPENDIX B

Geographical Distribution of CEI Data Call Respondents

The following heat maps (Figures B.1–B.4) represent the geographical distribution of personnel who responded to the CEI data call. Territories, such as American Samoa, were not included, as there were no respondents from these areas.

Figure B.1
ARNG Respondents with Any Level of Cyber Expertise

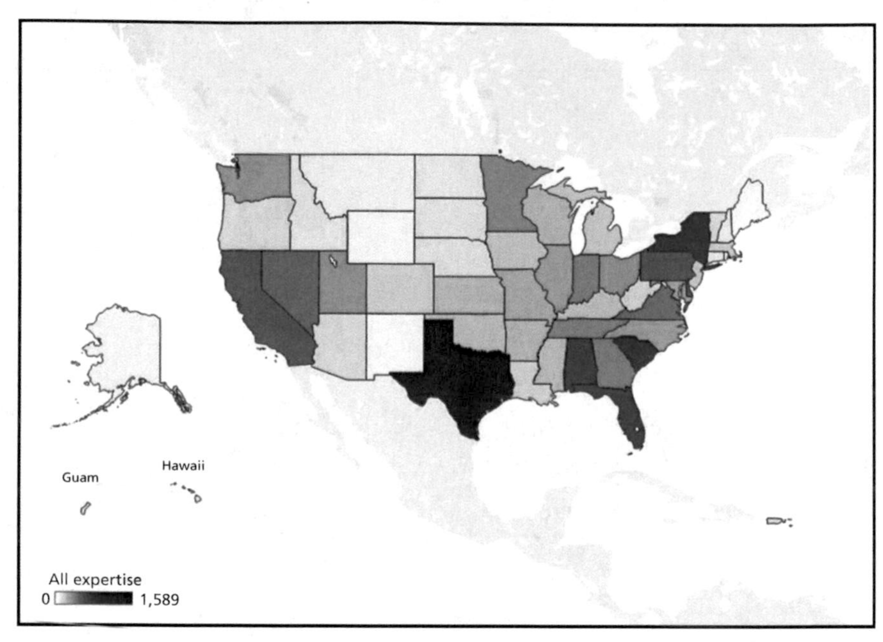

SOURCE: RAND Arroyo Center analysis of CEI data.
RAND *RR1490-B.1*

Figure B.2
ARNG Respondents with Maximum or Mid-Level Cyber Expertise

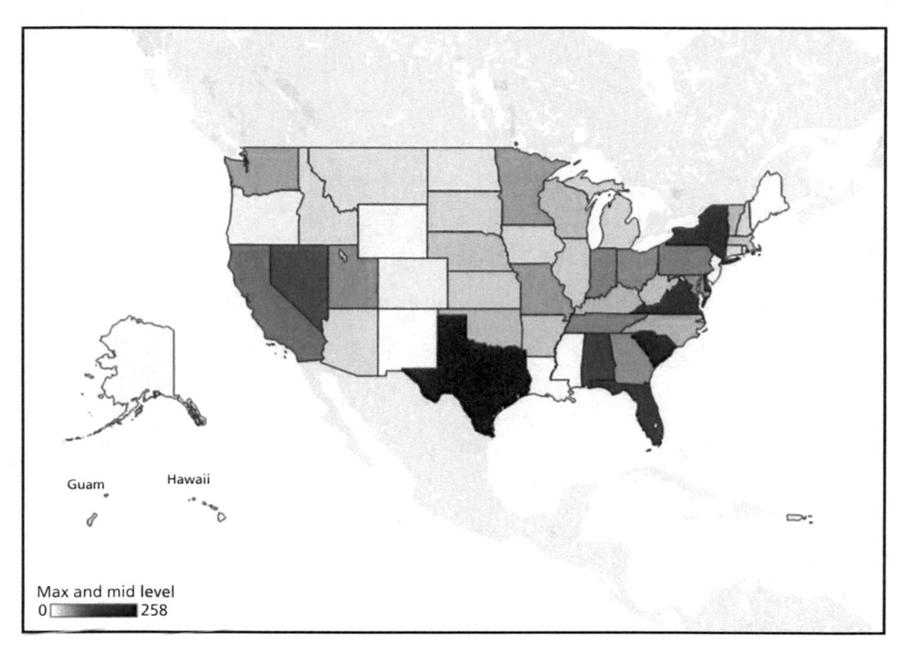

SOURCE: RAND Arroyo Center analysis of CEI data.
RAND *RR1490-B.2*

Figure B.3
USAR Respondents with Any Level of Cyber Expertise

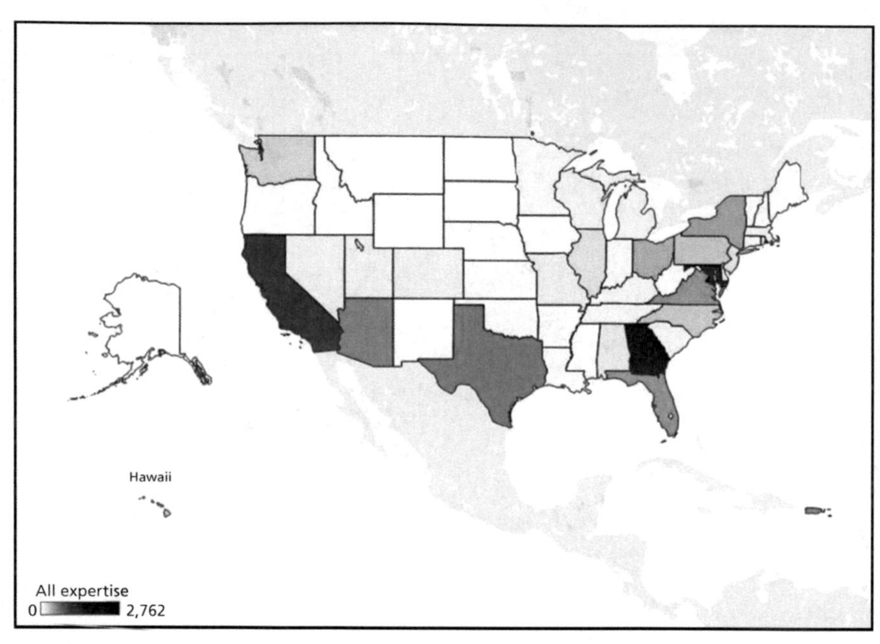

SOURCE: RAND Arroyo Center analysis of CEI data.
RAND *RR1490-B.3*

Figure B.4
USAR Respondents with Maximum or Mid-Level Cyber Expertise

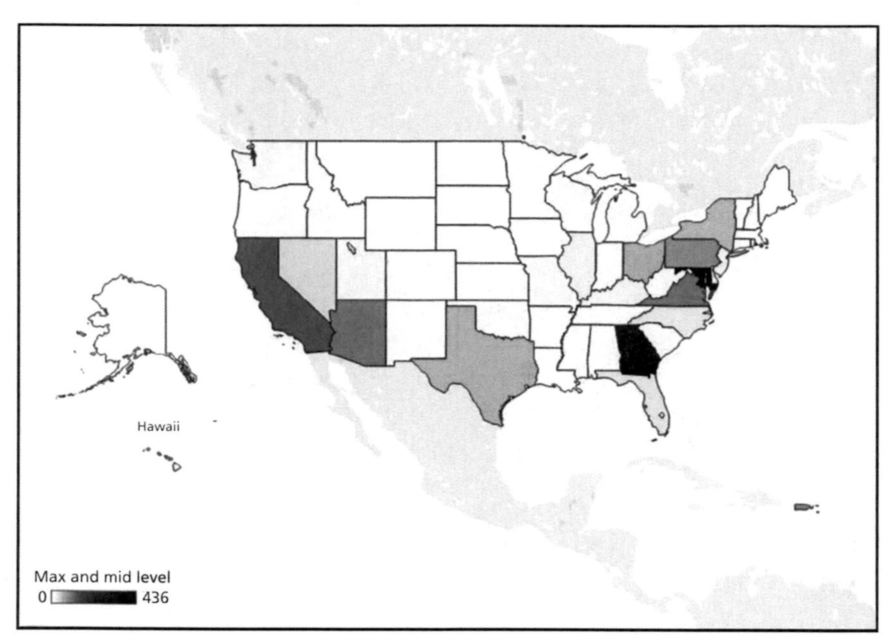

SOURCE: RAND Arroyo Center analysis of CEI data.

RAND *RR1490-B.4*

APPENDIX C

Select Army and Air Force Cyber Units

There has been cyber capability in the RC for over a decade.

The Army Reserve Cyber Operations Group

The Army has the ARIOC (recently renamed the ARCOG). This unit has a history of attracting personnel from the IT sector (including forensic experts from federal agencies). The history of the ARIOC is detailed by Good:

> In 2001, the Department of the Army requested the assistance of the Army Reserve to acquire a new range of support in the information operations field, with the broad requirement to take advantage of the high-technology skills of [reserve soldiers] already employed in the information technology industry. In response to this request, the Reserve created information operations units throughout the country. The units recruit [reserve soldiers] who can take the information technology expertise acquired from a civilian career in private industry and apply it in the context of their military service. By participating in the training developed by the Software Engineering Institute . . . these soldiers can further hone the technical skills and knowledge they bring to computer network defense operations. The majority of the Army Reserve Information Operations Command information operations soldiers are assigned to one of five information operations centers. The centers are strategically located in regions with a con-

> centration of crucial information technology resources: Washington, D.C.; Massachusetts; Pennsylvania; Texas; and California.[1]

Table C.1 describes various Air Force organizations involved in cyber missions.

Contributions to the Cyber Mission Force

There is already one Cyber Protection Team in the Army RC (with plans to grow to up to 11 in the near future), which supports a goal of further integrating reserve personnel into the Cyber Mission Force and Army Cyber Forces in particular.[2] An additional 400 personnel from

---

[1] Travis Good, "Army Reserve Trains for Information Assurance," *Signal*, January 2004.

[2] Sydney J. Freedberg, Jr., "The Army's Plan for Cyber, One Bright Spot in Its Budget," BreakingDefense.com, February 27, 2014; Darron Salzer, "Memorandum Establishes Commitment Between Guard, Army Cyber Command," *Defense Video and Imagery Distribution System*, June 6, 2014. Specifically, the memorandum of understanding describes an alignment of the Army Guard's 1636th Cyber Protection Team "under the command and control of ARCYBER [U.S. Army Cyber Command]."

Additionally, U.S. Cyber Commander Admiral Michael Rogers confirmed the important role of the RC during his confirmation hearing in March 2014 (U.S. Senate, Committee on Armed Services, *Nominations Before the Senate Armed Services Committee,* Second Session, 113th Congress, S. Hrg. 113–611, Washington, D.C., January 16; February 25; March 11; June 19; July 10, 17; December 2, 2014) when asked a question by Senator Richard Blumenthal:

> Senator BLUMENTHAL: The CYBERCOM Commander, General Alexander, frequently talked about the critical value of the National Guard as a resource and the role that it could play in expanding our military cyber warfare and defense capabilities. Do you agree with him and how would you define the value that the National Guard can bring to this effort?
>
> Admiral ROGERS. Yes, sir, I do agree. At the present, the Department as a matter of fact is in the process of doing the analysis right now to address that very question. . . . While the U.S. Navy does not have a Guard structure, the Reserve structure we use has been very effective for us. I have worked hard to try to apply it in my current duty.
>
> Senator BLUMENTHAL. And frequently those members of the Naval Reserve or of the National Guard, the Army National Guard or Air Force, bring capabilities, training, education, skills that are very valuable.
>
> Admiral ROGERS. Oh, yes, sir.

Table C.1
Select Air Force Cyber Organizations

Name	Organizational Affiliation	Size	Current and Past Air Force Roles and Missions	Reference
960th Cyberspace Operations Group[a]	Air Force Reserve Command		"Administrative control over 10 Reserve cyber organizations spread throughout the country"	Joyner, 2013
262nd NWS (within 194th Regional Support Wing)	Washington State Air Guard	100 airmen	Security assessment of state driver's license system and other state networks; study of security of industrial control systems; conduct cyber-emergency planning	"National Guardsmen: The New Front Line in Cyber Security," 2011; Matthews, 2012
299th Network Operations Security Squadron (NOSS) (within the 184th Intelligence Wing)	Kansas Air Guard		"Real-time network security to over 106,000 Air National Guard users"	184th Intelligence Wing, undated[b]
177th Information Warfare Aggressor Squadron (within 184th Intelligence Wing)	Kansas National Guard			Matthews, 2012
261st IOS (within 162nd Combat Communications Group)	California Air Guard	125	Can be called on by the governor to test the security of state networks	Matthews, 2012, 2014
175th NWS	Maryland National Guard	100-person squadron	"Cyber Hunter Squadrons . . . monitor networks for intrusions and unauthorized users [and to] develop countermeasures"; "Perform security assessments on state computer networks"	Matthews, 2012, 2014

Table C.1—continued

Name	Organizational Affiliation	Size	Current and Past Air Force Roles and Missions	Reference
102nd NWS	Rhode Island Air National Guard		"Monitor military networks for anomalies and suspicious activity and conduct cyber readiness inspections"	Matthews, 2012
166th NWS	Delaware National Guard		Support of the NSA	Matthews, 2014

[a] ARCOG, formerly known as the ARIOC, is a comparable organization within the Army.

[b] The NOSS provides real-time network security to more than 106,000 Air National Guard users at 300 locations. The NOSS manages network defense, generates an enterprise situational awareness picture, and manages network configuration. It also provides information assurance for all ANG networks, and serves as the network help desk for application and system issues throughout the entire ANG enterprise (184th Intelligence Wing, undated).

the Army Reserve are expected to be integrated into various units that support cyber operations.[3] This is based on the Total Army Analysis 16-20 "wedge."[4]

Table C.2 provides information about existing and planned Army RC units with cyber roles.

Table C.3 provides information on select organizations in other countries through which civilians support the cyber mission of their nations' militaries.

Exemplar: Role of Guard Units with Kansas Fusion Center

There is a fusion center located in Kansas called the Kansas Threat Integration Center, which was established in 2004.[5] Focused on counterterrorism, it is a joint operation of the Kansas Bureau of Investigation, the Kansas Highway Patrol, and the Kansas National Guard. The Kansas City Regional Terrorism Early Warning Group has an Interagency Analysis Center in Kansas City, Missouri, that is responsible for Leavenworth, Wyandotte, Johnson, and Miami counties.[6]

The fusion center cooperates with the National Guard via facilities and has network access and space for classified discussions with joint and interagency partners. The efforts in Kansas are viewed as productive by many.[7]

---

[3] See Jacqueline M. Hames, "Army Cyber Capabilities Increasing to Include Guard, Reserve," army.mil website, October 17, 2014.

[4] TAA (Total Army Analysis) is part of the force development process. "TAA is the process that takes us from the Army of today to the Army of the future. It requires a doctrinal basis and analysis; is based upon strategic guidance from above the Army; and involves threat analysis, specific scenarios, and an Army 'constrained' force" (Commonwealth Institute, *Total Army Analysis: Primer 2008*, Cambridge, Mass., undated).

[5] U.S. Department of Homeland Security, "Fusion Center Locations and Contact Information," web page, undated.

[6] "Top Secret America: A Washington Post Investigation—Kansas," *Washington Post*, 2015.

[7] See State of Kansas Adjutant General, *Annual Report 2014*, Topeka, Kan., 2014.

Table C.2
Current and Future Guard and Reserve Cyber Force and Operating Status

Component	Name	Number of Personnel	Mission Focus	Status
ARNG	1636th Cyber Protection Team	1 team	Title 10 mission, out of Laurel, Maryland	New unit, operating in Title 10 active duty status
ARNG	Cyber Network Defense Teams	432	State cybersecurity	Current unit, controlled by the state
ARNG	Virginia Data Processing Unit	174	Cyber operations support	Current unit, operated under Title 10
ARNG	M-Day Cyber Protection Team [a]	390 (10 teams)	DoD/U.S. Cyber Command/Army missions; surge capacity; critical infrastructure and key resource mission	Future concept, available for state missions and DSCA (DHS and FBI)
USAR	1st IO Troop Program Unit	70	Support and provide cyber and IO soldiers to cyber opposing forces	Current unit
USAR	335th Signal Command Det	88	Regional Cyber Center (NOSC)	Current unit in Kuwait
USAR	Military Intelligence Readiness Command	20	Intelligence support and analysis products for 780th MI Brigade	Current unit
USAR	ARIOC (renamed ARCOG)	308	DCO for Army networks	Current unit
USAR	ARCOG—CPTs	390 (10 teams)	DCO	Future concept
USAR	ARCOG – C2	79	Mission command for reserve CPTs	Future concept
USAR	ARISCO	84	Intelligence support to U.S. Army Cyber Command offensive teams	Future concept

Table C.2—continued

Component	Name	Number of Personnel	Mission Focus	Status
USAR	ARCC	50	Support to U.S. Army Cyber Command joint force headquarters	Future concept
USAR	CTSE	92	Opposing force support for exercises	Future concept
USAR	U.S. Cyber Command Army Reserve Element	23	Cyberspace planning and intelligence fusion	Current
USAR	DISA Army Reserve Element	95	Support DODIN mission	Current
USAR	DISA individual mobilization augmentees	16	Surge capacity for DISA	Current
USAR	U.S. Army Cyber Command individual mobilization augmentees	18	Surge capacity for U.S. Army Cyber Command	Current

SOURCE: Association of the United States Army, "RC Cyber Forces Concept," September 2, 2014.

[a] "M-Day" means "Mobilization Day," which stands for a traditional Army National Guard soldier who drills 48 Multiple Unit Training Assemblies and a 15-day annual training period a year as per statutory regulations" (Christopher Quick, *Creating a Total Army Cyber Force: How to Integrate the Reserve Component into the Cyber Fight*, Arlington, Va.: The Institute of Land Warfare, No. 103W, September 2014).

Table C.3
Summary of Various Cyber Volunteer Entities (Non-USA)

Entity	Status	Membership	Mission/ Responsibilities	Serving/Training Period
Cyber Defense Unit (Estonia)	Paramilitary	Voluntary; professionals	Cooperation enhancement between public and private sectors, knowledge sharing, awareness, member training	Depends on members
Warning Advice and Reporting program (WARP) (UK)	Legal entity of public law (or a nongovernmental organization)	Voluntary membership of entities of both public and private law	Incident reporting, early warning, expert advice	N/A
LIAG, 81st Signal Squadron (UK)	Military– Reserve (Territorial Army)	Voluntary (paid) professionals; picked up with careful examination of experience and skills	Providing information assurance communications, fixed telecoms expertise to British Army, Air Force, Navy	Serving at least 19 days a year
Local militias (China)	Organized at province/ municipality level around educational or research institutions	Voluntary (coercion from government, academic incentives for students); IT-savvy students, professionals	Varies from internal political issues (censorship monitoring) to attacking operations	Training for at least 4 weeks; Possibility to be involved in operations from home or educational institution
Patriot hactivists (proposed in India and Japan)	Informal entities/ institutionalized units	Voluntary (incentives and coercion from government); professionals	Attacking operations, response to attacks	N/A
Russian cyber volunteers	Informal entity	Voluntary; skill levels vary from amateurs to professionals	Attacking operations	Any time; online manuals

SOURCE: Basilaia, 2012.

[a] A *professional* in the framework of the table means a person with IT education and work experience in the IT sector.

Exemplar: The Air National Guard's Cyber Role

The U.S. Air Force has historically partnered AC units with RC units; not only in tables of equipment, but also in missioning. Tactics, techniques, and procedures (TTPs) and tools are shared; in fact, TTPs are not approved at AC units without review and coordination from their "assigned" RC counterpart. Individuals, sections, and units perform annual training at the location. With the proper facilities and equipping; both the Air Force and the Navy have been able to perform missions from RC locations.[8]

There is a perspective among some that the U.S. Air Force is better at Total Force integration—that is, making better use of their reserve components—relatively speaking. For this reason, it is worth noting that an early design for the 24th U.S. Air Force paired Air Guard and Air Force Reserve.

---

[8] Correspondence with research questionnaire participant.

APPENDIX D

How the Survey Was Conducted

Whom We Contacted

RAND Arroyo Center sent emails to at least one leader in each state and territory for the ARNG. RAND Arroyo Center also sent emails to the Commanding Generals of the major reserve component commands.

Table D.1
USAR Units Contacted

State	Unit
Alabama	87th Support Command
American Post Office Europe (APO AE)	7th Civil Support Command
California	79th Sustainment Support Command
Florida	Army Reserve Medical Command
Georgia	335th Signal Command (Theater)
Georgia	3rd Medical Command (Deployment Support)
Georgia	Army Reserve Careers Division
Hawaii	9th Mission Support Command
Illinois	416th Theater Engineer Command
Illinois	85th Support Command
Kentucky	11th Aviation Command
Kentucky	83rd USARRTC
Kentucky	84th Training Command
Louisiana	377th Theater Sustainment Command

Table D.1—continued

State	Unit
Maryland	200th Military Police Command
Maryland	USAR Legal Command
Mississippi	412th Theater Engineer Command
New Jersey	99th Regional Support Command
North Carolina	U.S. Army Civil Affairs & Psychological Operations Command (Airborne)
North Carolina	108th Training Command
Puerto Rico	1st Mission Support Command
South Carolina	81st Regional Support Command
Texas	75th Training Command
Utah	76th Operational Response Command
Utah	807th Medical Command (Deployment Support)
Virginia	Military Intelligence Readiness Command
Virginia	80th Training Command
Wisconsin	88th Regional Support Command

How the Link to the Survey Was Shared

RAND Arroyo Center provided a link to these commanding officers, most of them general officers, and asked these officers to share the link with their personnel. A sample letter is shown in Figure D.1.

Characterization of the Level of the Response Rate

We cannot determine the exact number of personnel that were asked to respond. Clearly, it was greater than the number of responses we received (~1,216). A small number of respondents (~20) indicated that they were not in the RC and those personnel were not included in the final tally.

Respondents were asked to assess their cyber skills and cyber-related skills. They were also asked to describe their military and/or civilian occupations and to supply other details, such as the number

Figure D.1
Sample Email Invitation

Good morning:

As part of a study being co-sponsored by OCAR and ARNG, I am surveying personnel across the Army National Guard and Reserve to understand the amount of cyber-related knowledge and skills that exists.

This short survey can be taken by going to this link:

https://www.randsurvey.org/skills/

Would you please help me by sharing this link with those uniformed personnel in your organization? This effort has been endorsed by MG Fogarty and that endorsement is attached to this email for your consideration. This survey is being offered to all uniformed reserve component personnel.

Please do not hesitate to contact me if you have any questions or comments.

Isaac R. Porche, Ph.D.

RAND *RR1490-D.1*

of relevant certificates (e.g., CISSP, Security+) they hold. Respondents were also asked to report on where they apply their related skills.

There was comparable variation in the metropolitan areas represented by the respondents, as shown in Figure D.3.

Figure D.2
Number of Respondents, by State

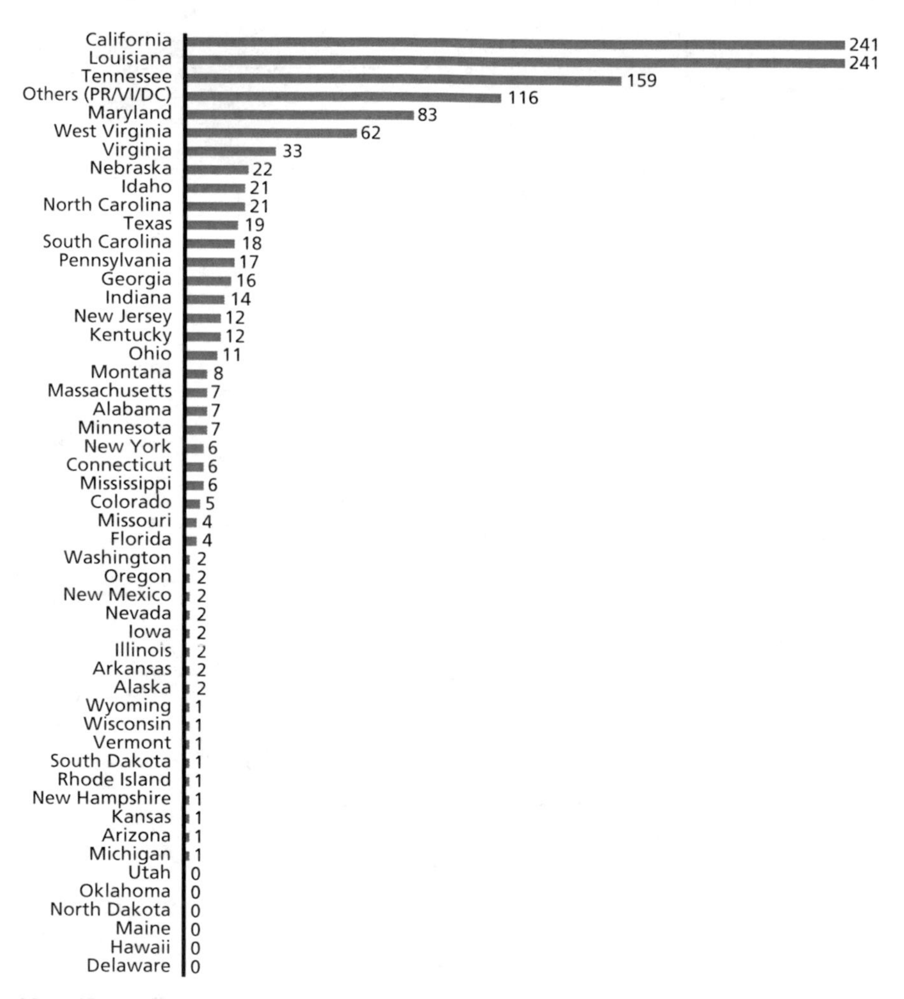

SOURCE: RAND Arroyo Center analysis of survey data.
RAND *RR1490-D.2*

Figure D.3
Top 20 Metropolitan Areas

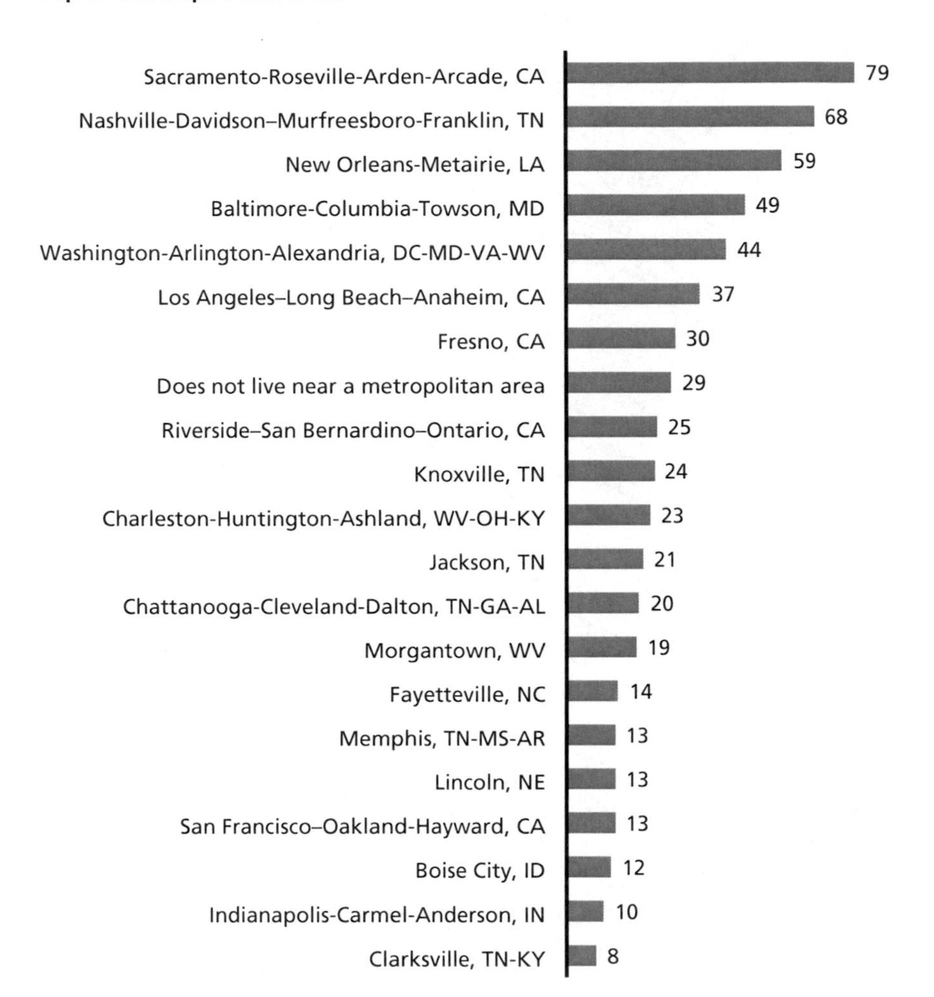

SOURCE: RAND Arroyo Center analysis of survey data.
RAND *RR1490-D.3*

Lists of Figures and Tables

Figures

Tables

Abbreviations

AC	active component
ARC	Air Reserve Component
ARCOG	Army Reserve Cyber Operations Group
ARIOC	Army Reserve Information Operations Command
ARNG	Army Reserve National Guard
ASI	additional skill identifier
BCT	brigade combat team
C2	command and control
C4ISR	command, control, communications, computers, intelligence, surveillance, and reconnaissance
CCIE	Cisco Certified Internet Expert
CCNA	Cisco Certified Network Associate
CCNP	Cisco Certified Network Professional
CEH	Certified Ethical Hacker
CEI	Civilian Employment Information
CERT	Computer Emergency Response Team
CISA	Certified Information Systems Auditor
CISSP	Certified Information Systems Security Professional

CNA	computer network attack
CND	computer network defense
CND-SP	Computer Network Defense Service Provider (program)
CompTIA	Computing Technology Industry Association
CONUS	continental United States
CPT	cyber protection team
DCO	defensive cyber operations
DHS	U.S. Department of Homeland Security
DISA	Defense Information Systems Agency
DoD	U.S. Department of Defense
DODIN	Department of Defense Information Network
DOTMLPF-P	doctrine, organization, training, materiel, leadership and education, personnel, facilities, and policy
DSCA	defense support of civil authorities
FFRDC	federally funded research and development center
ICS	industrial control system
IDA	Institute for Defense Analyses
infosec	information security
IO	information operations
IOS	information operations squadron
IR	information requirements
IT	information technology
JAG	Judge Advocate General

KSA	knowledge, skills, and abilities
LIAG	Land Information Assurance Group
MAVNI	Military Accessions Vital to the National Interest
MI	military intelligence
MOS	military occupational specialty
NOSC	Network Operations Support Center
NSA	National Security Agency
NWS	network warfare squadron
OCO	offensive cyber operations
PIR	priority information requirements
RC	reserve component
RCERT	Regional Commuter Emergency Response Team
ROTC	Reserve Officers' Training Corps
RSC	regional support command
SCADA	supervisory control and data acquisition
SSCP	Systems Security Certified Practitioner
SOC	standard occupational classification
STEM	science, technology, engineering, and mathematics
SWA	Southwest Asia
TTPs	tactics, techniques, and procedures
UIC	unit identification code
USAR	U.S. Army Reserve
UTC	unit type code
WEX	Work Experience File

References

184th Intelligence Wing, homepage, undated. As of March 20, 2017:
http://www.184iw.ang.af.mil/units/

Alexander, Keith B., "Statement of General Keith B. Alexander, Commander, United States Cyber Command," testimony to the Senate Armed Services Committee, Washington, D.C., March 13, 2013. As of March 20, 2017:
http://www.armed-services.senate.gov/imo/media/doc/Alexander%2003-12-13.pdf

"Army National Guard Stands Up Cyber Protection Teams," *Army Times*, March 1, 2013. As of August 27, 2015:
http://www.armytimes.com/story/military/guard-reserve/2015/03/01/army-national-guard-cyber-protection-teams/24003611/

Assistant Secretary of Defense for Networks and Information Integration/Department of Defense Chief Information Officer, *Information Assurance Workforce Improvement Program*, Washington, D.C.: U.S. Department of Defense, DoD 8570.01-M, January 24, 2012. As of March 20, 2017:
http://www.dtic.mil/whs/directives/corres/pdf/857001m.pdf

Associated Press, "North Korea Has 6,000-Strong Cyber Army, Says South Korea," *The Guardian*, January 6, 2015. As of March 20, 2017:
http://www.theguardian.com/world/2015/jan/06/north-korea-6000-strong-cyber-army-south-korea

Association of the United States Army, "RC Cyber Forces Concept," September 2, 2014.

Aubuchon, Kurt, "How Many Information Security Staff Do We Need?" *Infosec Island*, blog post, September 26, 2010. As of March 20, 2017:
http://infosecisland.com/blogview/8327-How-Many-Information-Security-Staff-Do-We-Need.html

Avgerinos, Thanassis, Sang Kil Cha, Brent Lim Tze Hao, and David Brumley, "AEG: Automatic Exploit Generation," Pittsburgh, Pa.: Carnegie Mellon University, undated. As of March 20, 2017:
http://security.ece.cmu.edu/aeg/aeg-current.pdf

Barnes, Thomas L., information assurance expert at the U.S. Army Cyber Center of Excellence, personal communication with the authors, December 24, 2015.

Basilaia, Mikheil, *Volunteers and Cyber Security: Options for Georgia*, Tallinn, Estonia: Tallinn University of Technology, 2012.

Borras, COL Aida T., Army action officer for this project, personal correspondence with the authors, November 9, 2015.

Boudreau, Todd, "Cyberspace Defense Technician (MOS 255S)," *Army Communicator*, Vol. 36, No. 1, Spring 2011a, pp. 35–39.

———, "Cyberspace Network Management Technicians (MOS 255N)," *Army Communicator*, Vol. 36, No. 1, Spring 2011b, pp. 30–34.

Bush, George W., *The National Security Strategy of the United States of America*, Washington, D.C.: The White House, March 2006.

Cardon, Edward C., "2014 Green Book: Army Cyber Command and Second Army," September 30, 2014. As of March 20, 2017:
http://www.army.mil/article/134857

———, "Operationalizing Cyberspace for the Services," testimony before the House Armed Services Committee Subcommittee on Emerging Threats and Capabilities, Washington, D.C., March 4, 2015. As of March 20, 2017:
http://docs.house.gov/meetings/AS/AS26/20150304/103093/HHRG-114-AS26-Wstate-CardonE-20150304.pdf

Central Reserve Headquarters Royal Signals, homepage, 2015. As of March 20, 2017:
http://www.army.mod.uk/signals/35432.aspx

Commonwealth Institute, *Total Army Analysis: Primer 2008*, Cambridge, Mass., undated. As of August 28, 2015:
http://www.comw.org/qdr/fulltext/08TAA.pdf

Computer Security Institute, *2010/2011 Computer Crime and Security Survey*, New York, 2011.

Cummings, Joanne, "Your Life in the Virtualized Future," *NetworkWorld*, July 26, 2004. As of March 20, 2017:
http://www.networkworld.com/you/2004/0726future.html?page=3

Deloitte, "Cybersecurity Workforce Planning: Building a Cyber-Savvy Federal Workforce for Tomorrow's Safety," 2012. As of March 2, 2017:
https://web.archive.org/web/20140802164420/http://www.deloitte.com/view/en_US/us/Industries/US-federal-government/bd32c62b5fb1a310VgnVCM2000003356f70aRCRD.htm

Deloitte and the National Association of State Chief Information Officers, *2012 Deloitte-NASCIO Cybersecurity Study: State Governments at Risk—A Call for Collaboration and Compliance*, 2012.

———, *2014 Deloitte-NASCIO Cybersecurity Study: State Governments at Risk—Time to Move Forward*, 2014.

Deloitte Touche Tohmatsu, *2003 Global Security Survey*, New York, May 2003.

Department of Defense Directive 8140.01, *Cyberspace Workforce Management*, Washington, D.C.: U.S. Department of Defense, August 11, 2015. As of March 20, 2017:
http://www.dtic.mil/whs/directives/corres/pdf/814001_2015_dodd.pdf

Department of Defense Instruction 8500.01, *Cybersecurity*, Washington, D.C.: U.S. Department of Defense, March 14, 2014. As of March 20, 2017:
http://www.dtic.mil/whs/directives/corres/pdf/850001_2014.pdf

Dias, Dennis P., *Partnering with Private Networks: The DoD Needs a Reserve Cyber Corps*, Carlisle, Pa.: U.S. Army War College, March 15, 2008. As of March 20, 2017:
http://www.dtic.mil/dtic/tr/fulltext/u2/a479001.pdf

Dion-Schwarz, Cynthia, manager of Cyber and Data Sciences Programs at the RAND Corporation, personal communication to the authors, December 10, 2015.

DoD—*See* U.S. Department of Defense.

Dudding, R. Wayne, former commander of the Army Information Operations Group, personal communication with the authors, January 31, 2016.

Ewing, Philip, "Ash Carter's Appeal to Silicon Valley: We're 'Cool' Too," *Politico*, April 23, 2015. As of March 20, 2017:
http://www.politico.com/story/2015/04/ash-carter-silicon-valley-appeal-117293

Ferrari, John G., "The Army Program FY16–20," Program Analysis and Evaluation Commission Brief, briefing slides, June 17, 2015. As of March 20, 2017:
http://www.ncfa.ncr.gov/sites/default/files/MG%20Ferrari%20PAE%20Brief%20
18%20Jun%202015.pdf

Field Manual 7-15, *Army Universal Task List*, Washinghton, D.C.: Headquarters Department of the Army, February 2009.

Fisher, Jeff L., and Brian Wisniewski, *Employment of Reserve Forces in the Army Cyber Structure*, Carlisle Barracks, Pa.: U.S. Army War College, May 2012.

Fort Gordon Public Affairs, "Army Cyber Branch Offers Soldiers New Challenges, Opportunities," army.mil website, November 24, 2014. As of March 20, 2017:
http://www.army.mil/article/138883/
Army_Cyber_branch_offers_Soldiers_new_challenges__opportunities/

Freedberg, Sydney J., Jr., "The Army's Plan for Cyber, One Bright Spot in Its Budget," BreakingDefense.com, February 27, 2014. As of March 20, 2017:
http://breakingdefense.com/2014/02/
the-armys-plan-for-cyber-one-bright-spot-in-its-budget/

Garamone, Jim, "Gates Lays Out Budget Recommendations," American Forces Press Service, April 6, 2009. As of March 2, 2017:
http://archive.defense.gov/news/newsarticle.aspx?id=53812

Good, Travis, "Army Reserve Trains for Information Assurance," *Signal*, January 2004. As of March 20, 2017:
http://www.afcea.org/content/?q=army-reserve-trains-information-assurance

Gowen, Lon D., *Predicting Staffing Sizes for Maintaining Computer Networking Infrastructures*, McLean, Va.: MITRE Corporation, 2000. As of August 13, 2015:
http://www.mitre.org/publications/technical-papers/
predicting-staffing-sizes-for-maintaining-computernetworking-infrastructures

Halla, David, U.S. Cyber Command/J7, personal communication with the authors, April 25, 2014.

Hames, Jacqueline M., "Army Cyber Capabilities Increasing to Include Guard, Reserve," army.mil website, October 17, 2014. As of March 20, 2017:
http://www.army.mil/article/136371/
Army_Cyber_capabilities_increasing_to_include_Guard__Reserve

Harris, Shane, "Pentagon Memo: U.S. Weapons Open to Cyberattacks," *The Daily Beast*, December 16, 2015. As of March 20, 2017:
http://www.thedailybeast.com/articles/2015/12/16/pentagon-memo-u-s-weapons-open-to-cyber-attacks.html

Hernandez, Rhett, former commander of U.S. Army Cyber Command, personal communication with the authors, December 15, 2015.

Hill, Kashmir, "The NSA Gives Birth to Start-Ups," *Forbes*, September 10, 2014.

Homan, Timothy R., "ADP Estimates Companies in U.S. Added 42,000 Jobs," *Bloomberg*, August 4, 2010.

Hosek, James, Michael G. Mattock, C. Christine Fair, Jennifer Kavanagh, Jennifer Sharp, and Mark E. Totten, *Attracting the Best: How the Military Competes for Information Technology Personnel*, Santa Monica, Calif.: RAND Corporation, MG-108-OSD, 2004. As of March 20, 2017:
http://www.rand.org/pubs/monographs/MG108.html

Hughes, Kelly J., "Ops for the WA Unit," email to the authors, June 13, 2014.

$(ICS)^2$ Inc., "DoD Fact Sheet," 2015.

The Intersector Project, "Cyber P3 to Build 'Network of Cyber Warriors,'" February 27, 2015. As of March 20, 2017:
http://intersector.com/cyber-p3-build-network-cyber-warriors-2/

"IT Security Staff Levels Are Declining," *Computer Economics*, August 2008. As of March 20, 2017
http://www.computereconomics.com/article.cfm?id=1384

Johnson, Nicole Blake, "The Air Force Has a Plan for Testing Cyber Aptitude," govloop.com, August 18, 2015. As of March 20, 2017: https://www.govloop.com/the-air-force-has-a-plan-for-testing-cyber-aptitude/

Joint Electronic Library, *Universal Joint Task List*, April 2015.

Joint Publication 1-02, *DoD Dictionary of Military and Associated Terms*, Washington, D.C.: Joint Chiefs of Staff, February 2017. As of March 2, 2017: http://www.dtic.mil/doctrine/new_pubs/dictionary.pdf

Joint Publication 3-12 (R), *Cyberspace Operations*, Washington, D.C.: Joint Chiefs of Staff, February 5, 2013. As of March 20, 2017: http://www.dtic.mil/doctrine/new_pubs/jp3_12R.pdf

Jontz, Sandra, "Uniting Cyber Defenses," *SIGNAL*, October 1, 2015. As of March 20, 2017: http://www.afcea.org/content/?q=Article-uniting-cyber-defenses

Joyner, Bo, "Reserve Activates Cyberspace Operations Group," *U.S. Air Force News*, March 1, 2013. As of March 20. 2017: http://www.af.mil/News/Article-Display/Article/109635/reserve-activates-cyberspace-operations-group/

JP—*See* Joint Publication.

Kansas Air National Guard, "184th Intelligence Wing," undated.

Kelsall, Chris, "DON IT Workforce," briefing to integrated product team, September 2009.

Kvavik, Robert B., John Voloudakis, Judith B. Caruso, Richard N. Katz, Paula King, and Judith A. Pirani, *Information Technology Security: Governance, Strategy, and Practice in Higher Education*, Louisville, Colo.: Educause Center for Applied Research, 2003.

Lawrence, Dune, "The U.S. Government Wants 6,000 New 'Cyberwarriors' by 2016," *Bloomberg Business*, April 15, 2014. As of March 20, 2017: http://www.businessweek.com/articles/2014-04-15/uncle-sam-wants-cyber-warriors-but-can-he-compete

Lester, Michael, Paul Gross, Carrie McLeish, and Bryan Rude, "Connect: Cyber Support to Joint Information Environment (JIE)," briefing presented at AFCEA TechNet, Augusta, Ga., September 9, 2014. As of March 20, 2017: http://www.afcea.org/events/augusta/14/documents/AFCEATechNetWOPanelFINAL20140908.pdf

Levine, Daniel B., and Stanley A. Horowitz, *A New Approach to Force-Mix Analysis: A Case Study Comparing Air Force Active and Reserve Forces Conducting Cyber Missions*, Alexandria, Va.: Institute for Defense Analyses, September 2013. As of March 20, 2017: http://www.dtic.mil/cgi-bin/GetTRDoc?AD=ADA591198

Libicki, Martin C., David Senty, and Julia Pollak, *Hackers Wanted: An Examination of the Cybersecurity Labor Market*, Santa Monica, Calif.: RAND Corporation, RR-430, 2014. As of March 20, 2017:
http://www.rand.org/pubs/research_reports/RR430

LinkedIn, "About LinkedIn," web page, July 2015. As of March 20, 2017:
https://press.linkedin.com/about-linkedin

Mandiant, *2013 Threat Report*, Alexandria, Va., 2013.

Mansharof, Yossi, *Iran's Cyber War: Hackers in Service of the Regime; IRGC Claims Iran Can Hack Enemy's Advanced Weapons Systems; Iranian Army Official: 'The Cyber Arena Is Actually the Arena of the Hissen Imam,'* Washington, D.C.: The Middle East Media Research Institute, Inquiry and Analysis Series Report No. 1012, August 25, 2013. As of March 20, 2017:
http://www.memri.org/report/en/print7371.htm#_edn13

Matthews, William, "Growth Mission," *National Guard Magazine*, Vol. 66, No. 6, June 2012, pp. 22–25. As of March 20, 2017:
http://www.nationalguardmagazine.com/publication/?i=114111

———, "Cyber Uncertainty," *National Guard Magazine*, July 2014. As of April 8, 2017:
http://nationalguardmagazine.com/article/Cyber_Uncertainty/1764536/218066/article.html

McCaney, Kevin, "Army's New Cyber Branch Looking to Recruit Talent," *Defense Systems*, December 11, 2014. As of March 20, 2017:
http://defensesystems.com/articles/2014/12/11/army-cyber-branch-new-career-field.aspx

McConnell, John M., untitled remarks delivered at the Spring 2015 Senator Christopher S. "Kit" Bond lecture series, YouTube.com, March 12, 2015. As of March 20, 2017:
https://www.youtube.com/watch?v=_RPT9pAVUsY

McKinley, Craig, *The National Guard: A Great Value Today and in the Future*, Washington, D.C.: National Guard Bureau, 2011.

Miller, Drew, Daniel B. Levine, and Stanley A. Horowitz, *A New Approach to Force-Mix Analysis: A Case Study Comparing Air Force Active and Reserve Forces Conducting Cyber Missions*, Alexandria, Va.: Institute for Defense Analyses, P-4986, September 2013.

Mitchell, Robert L., "Enterprise Linux? Not So Fast," *ComputerWorld*, January 19, 2009. As of March 20, 2017:
http://www.computerworld.com/article/2550718/operating-systems/enterprise-linux--not-so-fast-.html

Moore, Jack, "In Fierce Battle for Cyber Talent, Even NSA Struggles to Keep Elites on Staff," nextgov.com, April 14, 2015. As of March 20, 2017: http://www.nextgov.com/cybersecurity/2015/04/fierce-battle-cyber-talent-even-nsa-struggles-keep-elites-staff/110158/

———, "The NSA's Fight to Keep Its Best Hackers," *DefenseOne*, March 20, 2017. As of August 26, 2015: http://www.defenseone.com/management/2015/04/nsas-fight-keep-its-best-hackers/110401/

El Nasser, Haya, "Geek Chic: 'Brogrammer?' Now, That's Hot," *USA Today*, April 12, 2012. As of March 20, 2017: http://usatoday30.usatoday.com/tech/news/story/2012-04-10/techie-geeks-cool/54160750/1

National Cybersecurity Education Office, *2012 Information Technology Workforce Assessment for Cybersecurity (ITWAC): Summary Report*, Washington, D.C.: U.S. Department of Homeland Security, March 14, 2013. As of April 18, 2017: https://niccs.us-cert.gov/sites/default/files/documents/pdf/itwac_summary_report_04_01_2013.pdf?trackDocs=itwac_summary_report_04_01_2013.pdf

"National Guardsmen: The New Front Line in Cyber Security," *Homeland Security News Wire*, December 19, 2011. As of March 20, 2017: http://www.homelandsecuritynewswire.com/dr20111219-national-guardsmen-the-new-front-line-in-cybersecurity

National Initiative for Cybersecurity Education, *The National Cybersecurity Workforce Framework*, Washington, D.C.: National Institute of Standards and Technology, 2013. As of March 20, 2017: http://csrc.nist.gov/nice/framework/national_cybersecurity_workforce_framework_03_2013_version1_0_for_printing.pdf

National Research Council, Office of Scientific and Engineering Personnel, *Building a Workforce for the Information Economy*, Washington, D.C.: The National Academies Press, 2001.

National Security Agency, *National Information Systems Security (INFOSEC) Glossary*, Washington, D.C., NSTISSI No. 4009, September 2000. As of March 20, 2017: http://handle.dtic.mil/100.2/ADA433929

Navy Cyber ZBR Task Force, *Navy Cyber Workforce Zero-Based Review*, April 2012.

NSA—*See* National Security Agency.

Office of the Secretary of Defense, *Cyber Mission Analysis: Mission Analysis for Cyber Operations of Department of Defense* [also known as "The Section 933 report"], Washington, D.C., August 21, 2014, not available to the general public.

Partnership for Public Service and Booz Allen Hamilton, *Cyber IN-Security: Strengthening the Federal Cybersecurity Workforce*, Washington, D.C., July 2009. As of March 20, 2017:
https://ourpublicservice.org/publications/download.php?id=121

Paul, Christopher, Isaac R. Porche III, and Elliot Axelband, *The Other Quiet Professionals: Lessons for Future Cyber Forces from the Evolution of Special Forces*, Santa Monica, Calif.: RAND Corporation, RR-780-A, 2014. As of March 20, 2017:
http://www.rand.org/pubs/research_reports/RR780

Petitt, Karen, "Cyberspace Career Fields, Training Paths, Badge Proposed," news release, Air Force Cyber Command (Provisional) Public Affairs, July 10, 2008.

Pirani, Judith A., *High Stakes: Strategies for Optimal IT Security Staffing*, Louisville, Colo.: Educause Center for Applied Research, Vol. 2004, No. 6, March 16, 2004. As of March 20, 2017:
http://net.educause.edu/ir/library/pdf/ERB0406.pdf

Ponemon Institute LLC, *Understaffed and at Risk: Today's IT Security Department*, Traverse City, Mich.: Ponemon Institute LLC, February 2014. As of March 2, 2017:
http://www.hp.com/hpinfo/newsroom/press_kits/2014/RSAConference2014/Ponemon_IT_Security_Jobs_Report.pdf

Porche, Isaac R., III, Christopher Paul, Michael York, Chad C. Serena, Jerry M. Sollinger, Elliot Axelband, Endy M. Daehner, and Bruce J. Held, *Redefining Information Warfare Boundaries for an Army in a Wireless World*, Santa Monica, Calif.: RAND Corporation, MG-1113-A, 2013. As of March 20, 2017:
http://www.rand.org/pubs/monographs/MG1113

Porst, Rolf, and Klaus Zeifang, "A Description of the German General Social Survey Test-Retest Study and Report on the Stability of Sociodemographic Variables," *Sociological Methods and Research*, Vol. 15, No. 3, February 1987, pp. 177–218.

Public Law 113-66, National Defense Authorization Act for Fiscal Year 2014, December 26, 2013.

Quick, Christopher, *Creating a Total Army Cyber Force: How to Integrate the Reserve Component into the Cyber Fight*, Arlington, Va.: The Institute of Land Warfare, September 2014, No. 103W. As of March 20, 2017:
https://www.ausa.org/publications/creating-total-army-cyber-force-how-integrate-reserve-component-cyber-fight

Quorum Technologies, Inc., "Case Study: Alameda County Medical Center," 2008.

Reed, John, "Unit 8200: Israel's Cyber Spy Agency," *Financial Times*, July 10, 2015. As of March 20, 2017:
http://www.ft.com/cms/s/2/69f150da-25b8-11e5-bd83-71cb60e8f08c.html

Reserve Forces Policy Board, *Report of the Reserve Forces Policy Board on Department of Defense Cyber Approach: Use of the National Guard and Reserve in the Cyber Mission Force*, RFPB Report FY14-03, August 18, 2014.

Richardson, Brian, "Improve Staffing Ratios," *ZDNet*, February 11, 2002.

Riley, Michael, Ben Elgin, Dune Lawrence, and Carol Mattack, "Missed Alarms and 40 Million Stolen Credit Card Numbers: How Target Blew It," *Bloomberg Business*, March 13, 2014. As of March 20, 2017:
http://www.bloomberg.com/bw/articles/2014-03-13/target-missed-alarms-in-epic-hack-of-credit-card-data

Roach, Brian, "3 Reasons Software-Defined Networking Is Streamlining DoD IT," *Defense Systems*, April 14, 2015. As of March 20, 2017:
http://defensesystems.com/articles/2015/04/14/comment-sdn-software-defined-networking-dod.aspx

Robbert, Al, James H. Bigelow, John E. Boon, Jr., Lisa M. Harrington, Michael McGee, S. Craig Moore, Daniel M. Norton, and William W. Taylor, *Suitability of Missions for the Air Force Reserve Components*, Santa Monica, Calif.: RAND Corporation, RR-429-AF, 2014. As of March 20, 2017:
http://www.rand.org/pubs/research_reports/RR429.html

Robbert, Al, Lisa M. Harrington, Tara L. Terry, and Hugh G. Massey, *Air Force Manpower Requirements and Component Mix: A Focus on Agile Combat Support*, Santa Monica, Calif.: RAND Corporation, RR-617-AF, 2014. As of March 14, 2017:
http://www.rand.org/pubs/research_reports/RR617.html

Ruderman, David, "Army Offers Selective Retention Bonuses to Retain Enlisted Cyber Warriors," army.mil website, May 29, 2015. As of March 20, 2017:
http://www.army.mil/article/149561

Salzer, Darron, "Memorandum Establishes Commitment Between Guard, Army Cyber Command," *Defense Video and Imagery Distribution System*, June 6, 2014. As of March 20, 2017:
http://www.dvidshub.net/news/132364/memorandum-establishes-commitment-between-guard-army-cyber-command#.U5cA4fldX_E

SANS Institute, *Cybersecurity Professional Trends: A SANS Survey*, May 2014.

Schmidt, Lara, Caolionn O'Connell, Hirokazu Miyake, Akhil R. Shah, Joshua Baron, Geof Nieboer, Rose Jourdan, David Senty, Zev Winkelman, Louise Taggart, Susanne Sondergaard, and Neil Robinson, *Cyber Practices: What Can the U.S. Air Force Learn from the Commercial Sector?* Santa Monica, Calif.: RAND Corporation, RR-847-AF, 2015. As of April 3, 2017:
https://www.rand.org/pubs/research_reports/RR847.html

Scott, Lynn M., Raymond E. Conley, Richard Mesic, Edward O'Connell, and Darren D. Medlin, *Human Capital Management for the USAF Cyber Force*, Santa Monica, Calif.: RAND Corporation, DB-579-AF, 2012. As of March 20, 2017: http://www.rand.org/pubs/documented_briefings/DB579.html

Solivan, Douglas A., Sr., "Communications-Electronics Cyber Training Range Launches," *Fort Gordon Globe*, July 10, 2015. As of March 20, 2017: http://www.fortgordonglobe.com/news/2015-07-10/News_Update/CommunicationsElectronics_cyber_training_range_lau.html

State of Kansas Adjutant General, *Annual Report 2014*, Topeka, Kan., 2014.

Suby, Michael, *The 2013 (ISC)² Global Information Security Workforce Study*, Mountain View, Calif.: Frost & Sullivan, 2013.

Suby, Michael, and Frank Dickson, *The 2015 (ISC)² Global Information Security Workforce Study*, Mountain View, Calif.: Frost & Sullivan, 2015.

Takalkar, Pradnya, Gordon Waugh, and Theodore Micceri, "A Search for Truth in Student Responses to Selected Survey Items," paper presented at AIR Forum, Chicago, Ill., May 15–19, 1993.

Talley, LTG Jeffrey W., *The 2016 Posture of the United States Army Reserve: A Global Operational Reserve Force*, submitted to the U.S. House of Representatives Appropriations Committee, March 22, 2016. As of March 17, 2017: http://www.usar.army.mil/Portals/98/Documents/resources_docs/2016ArmyReservePostureStatement.pdf

Tan, Michelle "Army Activates Its First Cyber Protection Brigade," *Army Times*, September 9, 2014. As of March 20, 2017: https://www.armytimes.com/story/military/tech/2014/09/09/army-activates-its-first-cyber-protection-brigade-/15352367/

Tice, Jim, "Officers Can Apply to Go Cyber in Voluntary Transfer Program," *Army Times*, October 6, 2014. As of March 20, 2017: https://www.armytimes.com/story/military/careers/army/officer/2014/10/08/officers-can-apply-to-go-cyber-in-voluntary-transfer-program/16925695/

———, "Reclassification Cash for Sergeants, Staff Sergeants," *Army Times*, May 6, 2015a. As of March 20, 2017: http://www.armytimes.com/story/military/careers/army/2015/05/06/reclassify-jobs-sergeants-staffsergeants/26585299/

———, "Staffing Goal for Cyber Branch Totals Nearly 1,300 Officers, Enlisted Soldiers," *Army Times*, June 15, 2015b. As of March 20, 2017: http://www.armytimes.com/story/military/2015/06/15/cyber-transfer-panels-and-reclassification-actions/71060716/

Tipton, Harold F., and Micki Krause, *Information Security Management Handbook*, 6th edition, Boca Raton, Fla.: Auerbach Publications, 2007.

Trippe, D. Matthew, Karen O. Moriarty, Teresa L. Russell, Thomas R. Caretta, and Adam S. Beatty, "Development of a Cyber/Information Technology Knowledge Test of Military Enlisted Technical Training Qualification," *Military Psychology*, Vol. 26, No. 3, 2014, pp. 182–198.

"Top Secret America: A Washington Post Investigation—Kansas," *Washington Post*, 2015. As of March 28, 2016:
http://projects.washingtonpost.com/top-secret-america/states/kansas/

U.S. Army, "Organization," web page, undated-a. As of March 20, 2017:
http://www.army.mil/info/organization/

———, "U.S. Army STEM Experience," web page, undated-b. As of March 17, 2017:
https://web.archive.org/web/20161230182746/http://www.goarmy.com/events/us-army-stem-experience.html

———, "Military Leave for Inactive Duty Training," June 6, 2006. As of April 3, 2017:
http://cpol.army.mil/library/permiss/5017a.html

———, *Army Reserve Medicine*, RPI 720 FS, May 2011. As of March 20, 2017:
https://www.goarmy.com/content/dam/goarmy/downloaded_assets/pdfs/amedd/RPI%20720%20FS%20Army%20Reserve%20Medicine%20Sep%2011%20%20LowRes.pdf

U.S. Army Cyber Command, "Summit Brings Senior Cyber Leaders Together to Share Total Army Opportunities, Solutions," army.mil website, January 5, 2016. As of March 20, 2017:
http://www.army.mil/article/160551/Summit_brings_senior_cyber_leaders_together_to_share_Total_Army_opportunities__solutions/

U.S. Army Reserve, "USAR Cyber P3," army.mil website, undated.

U.S. Bureau of Labor Statistics, "Employment Projections," web page, undated-a. As of September 3, 2015:
http://www.bls.gov/emp/

———, "Standard Occupational Classification," web page, undated-b. As of March 20, 2017:
http://www.bls.gov/soc/

———, "Benchmark Information, Comparison of All Employees, Seasonally Adjusted," 2014a. As of March 20, 2017:
http://www.bls.gov/ces/

———, "Information Security Analysts: Summary," January 8, 2014b. As of March 20, 2017:
http://www.bls.gov/ooh/computer-and-information-technology/information-security-analysts.htm

U.S. Department of Defense, *2013 Demographics: Profile of the Military Community*, Washington, D.C., undated. As of March 20, 2017:
http://www.militaryonesource.mil/12038/MOS/Reports/2013-Demographics-Report.pdf

———, *Cyber Operations Personnel Report*, Washington, D.C.: April 2011.

———, *The DoD Cyber Strategy*, Washington, D.C., April 2015. As of March 20, 2017:
http://www.defense.gov/Portals/1/features/2015/0415_cyber-strategy/Final_2015_DoD_CYBER_STRATEGY_for_web.pdf

U.S. Department of Homeland Security, "Fusion Center Locations and Contact Information," web page, undated. As of March 2, 2017:
https://www.dhs.gov/fusion-center-locations-and-contact-information

———, *Cyber Skills Task Force Report*, Washington, D.C., Fall 2012.

U.S. Government Accountability Office, *Military Personnel: Additional Actions Needed to Improve Oversight of Reserve Employment Issues*, Washington, D.C., GAO-07-259, February 2007. As of March 20, 2017:
http://www.gao.gov/assets/260/256366.pdf

———, *Cybersecurity Human Capital: Initiatives Need Better Planning and Coordination*, Washington, D.C.: GAO-12-8, November 2011.

U.S. Senate, Committee on Armed Services, *Nominations Before the Senate Armed Services Committee, Second Session, 113th Congress*, S. Hrg. 113–611, Washington, D.C., January 16; February 25; March 11; June 19; July 10, 17; December 2, 2014.

Vergun, David, "Cyber Chief: Army Cyber Force Growing 'Exponentially,'" army.mil website, March 5, 2015. As of March 20, 2017:
http://www.army.mil/article/143948/Cyber_chief__Army_cyber_force_growing__exponentially

Vijayan, Jaikumar, "Major Companies, Like Target, Often Fail to Act on Malware Alerts," *Computerworld*, March 14, 2014. As of March 20, 2017:
http://www.computerworld.com/article/2488641/malware-vulnerabilities/major-companies--like-target--often-fail-to-act-on-malware-alerts.html

Vostrom Holdings, Inc., "Staffing the Information Security Organization: Rationalizing the Staffing Requirements of a Reliable INFOSEC Team," undated. As of March 20, 2017:
https://vostrom.com/get/InfoSec_Staffing.pdf

Wikipedia, "National Guard of the United States," August 2015. As of March 20, 2017:
https://en.wikipedia.org/wiki/National_Guard_of_the_United_States

Wood, Charles Cresson, *Information Security and Data Privacy Staffing Levels: Benchmarking the Information Security Function*, Houston, Tex.: Information Shield, January 2012. As of March 20, 2017:
http://www.informationshield.com/papers/2011SecurityPrivacyStaffingSurvey.pdf